ADDITIONAL PRAISE FOR *SEX AT DAWN*

"Ideas like [these] might do more to save marriage than anything else in today's social-theory landscape. Seriously."

—Amy L. Keyishian, *AOL*

"*Sex at Dawn* is an absolute must-read because this is the first book about sex that's as fun to read as it is to . . . you know, do it."

—Moses Ma, editor, *Tantric News*

"*Sex at Dawn* is a page-turner. It's like a novel. You can't put it down—it is so much fun to read. . . . It's a fascinating book. I felt like this book was written for me!"

—Susie Bright, legendary sex-positive feminist and author

"I have not read such delightful, convincing, and readable science writing since the dearly lamented Stephen Jay Gould. This book is funny, absorbing, clear-eyed, and deeply anti-patriarchal in a way that feels incidental to the facts rather than rising from any agenda—which I find utterly, gleefully vindicating and deeply satisfying. . . . It made me laugh out loud every ten pag[illegible]tantly, this book made me proud to be [illegible]

[illegible]luct Press

"*Sex at Dawn* is the best, mo[illegible], yet ultimately inspirational book on the ev[illegible]ure of human sexuality that's out there!"

—Susan Block, Ph.D., sexologist, author, radio and television host

"An exciting book. . . . Whether people agree with it or not: these are issues that will need debating over and over before we will arrive at a resolution."

—Frans de Waal, Ph.D., author of *The Age of Empathy: Nature's Lessons for a Kinder Society*

"Turns everything you thought you knew about sex on its head. Funny, engaging, and superbly written, this book explores the science behind what many of us suspected all along: human beings are not naturally monogamous." —Julie Holland, M.D., author of *Weekends at Bellevue*

"*Sex at Dawn* manages to be both enormously erudite and wildly entertaining—even, frequently, hilarious. Ryan and Jethá slip effortlessly across millions of years, from the savanna of prehistoric Africa to the contemporary bedroom, presenting cutting-edge research with clarity and wit." —Tony Perrottet, author of *The Sinner's Grand Tour*

"This is a provocative, entertaining, and pioneering book. I learned a lot from it and recommend it highly."

—Andrew Weil, MD, Program Director,
Arizona Center for Integrative Medicine

"*Sex at Dawn* is not a tome on why people should cheat on their partners. Think of it as a new, wide-ranging sampling of research and ideas to get us to rethink our notion of human beings as sexual beings. . . . It helps put the 'human' back in 'human sexuality.' As suitable for an open-minded book club as the veteran sex therapist seeking new ways to rethink common challenges faced in clinical practice." —Eric Marlowe Garrison, *Contemporary Sexuality*

SEX

AT

DAWN

SEX AT DAWN

HOW WE MATE, WHY WE STRAY, AND WHAT IT MEANS FOR MODERN RELATIONSHIPS

Christopher Ryan, PhD,

and Cacilda Jethá, MD

HARPER PERENNIAL

NEW YORK • LONDON • TORONTO • SYDNEY • NEW DELHI • AUCKLAND

HARPER PERENNIAL

A hardcover edition of this book was published in 2010 by HarperCollins Publishers.

P.S.™ is a trademark of HarperCollins publishers.

FIRST HARPER PERENNIAL EDITION PUBLISHED 2011.

Designed by Joy O'Meara

The Library of Congress has catalogued the hardcover edition as follows:
Ryan, Christopher
Sex at dawn : the prehistoric origins of modern sexuality / Christopher Ryan and Cacilda Jethá.—1st ed.
p. cm.
Summary: "A controversial, idea-driven book that challenges everything you know about sex, marriage, family, and society."—Provided by publisher
Includes bibliographical references and index.
ISBN 978-0-06-170780-3 (hardback)
1. Sex. 2. Sex—History. 3. Sex customs. 4. Marriage. I. Jethá, Cacilda. II. Title.
HQ12.R93 2010
306.7—dc22

2009045457

ISBN 978-0-06-170781-0 (pbk.)

19 20 21 OV/LSC 40 39 38

Your children are not your children.
They are the sons and daughters
of life's longing for itself.

KAHLIL GIBRAN

CONTENTS

PREFACE

A Primate Meets His Match
(A note from one of the authors)

Nature, Mr. Allnut, is what we are put in this world to rise above.

KATHARINE HEPBURN,
as Miss Rose Sayer, in *The African Queen*

One muggy afternoon in 1988, some local men were selling peanuts at the entrance to the botanical gardens in Penang, Malaysia. I'd come with my girlfriend, Ana, to walk off a big lunch. Sensing our confusion, the men explained that the peanuts weren't for us, but to feed irresistibly cute baby monkeys like those we hadn't yet noticed rolling around on the grass nearby. We bought a few bags.

We soon came to a little guy hanging by his tail right over the path. His oh-so-human eyes focused imploringly on the bag of nuts in Ana's hand. We were standing there cooing like teenage girls in a kitten shop when the underbrush exploded in a sudden simian strike. A full-grown monkey flashed past me, bounced off Ana, and was gone—along with the nuts. Ana's hand was bleeding where he'd scratched her. We were stunned, trembling, silent. There'd been no time to scream.

After a few minutes, when the adrenaline had finally begun to ebb, my fear curdled into loathing. I felt betrayed in a way I never had before. Along with our nuts went precious assumptions about the purity of nature, of evil as a uniquely human affliction. A line had been crossed. I wasn't just angry; I was philosophically offended.

I felt something changing inside me. My chest seemed to swell, my shoulders to broaden. My arms felt stronger; my eyesight sharpened. I felt like Popeye after a can of spinach. I glared into the underbrush like the heavyweight primate I now knew myself to be. I'd take no more abuse from these lightweights.

I'd been traveling in Asia long enough to know that monkeys there are nothing like their trombone-playing, tambourine-banging cousins I'd seen on TV as a kid. Free-living Asian primates possess a characteristic I found shocking and confusing the first time I saw it: self-respect. If you make the mistake of holding the gaze of a street monkey in India, Nepal, or Malaysia, you'll find you're facing a belligerently intelligent creature whose expression says, with a Robert DeNiro–like scowl, "What the hell are *you* looking at? You wanna piece of me?" Forget about putting one of these guys in a little red vest.

It wasn't long before we came to another imploring, furry face hanging upside down from a tree in the middle of a clearing. Ana was ready to forgive and forget. Though I was fully hardened against cuteness of any kind, I agreed to give her the remaining bag of nuts. We seemed safely distant from underbrush from which an ambush could be launched. But as I pulled the bag out of my sweat-soaked pocket, its cellophane rustle must have rung through the jungle like a clanging dinner bell.

In a heartbeat, a large, arrogant-looking brute appeared at the edge of the clearing, about twenty yards away. He gazed at us, considering the situation, sizing me up. His exaggerated yawn seemed calculated to dismiss and threaten me simultaneously: a long, slow display of his fangs. Determined to fill any power vacuum without delay, I picked up a small branch and tossed it casually in his direction, making the point that these nuts were definitely not for him and that I was not to be trifled with. He watched the branch land a few feet in front of him, not moving a muscle. Then his forehead briefly crinkled in eerily emotional thought, as if I'd hurt his feelings. He looked up at me, straight into my eyes. His expression held no hint of fear, respect, or humor.

As if shot from a cannon, he leapt over the branch I'd tossed, long yellow dagger fangs bared, shrieking, charging straight at me.

Caught between the attacking beast and my terrified girlfriend, I understood for the first time what it would really mean to have a "monkey on your back." I felt something snap in my mind. I lost it. In movement quicker than thought, my arms flew open, my legs flexed into a wrestler's crouch, and my own coffee-stained, orthodontia-corrected teeth were bared with a wild shriek. I was helplessly launched into a hopping-mad, saliva-spraying dominance display of my own.

I was as surprised as he was. He pulled up and stared at me for a second or two before slowly backing away. This time, though, I'm pretty sure I saw a hint of laughter in his eyes.

Above nature? Not a chance. Take it from Mr. Allnut.

INTRODUCTION

Another Well-Intentioned Inquisition

Forget what you've heard about human beings having descended from the apes. We didn't descend from apes. We *are* apes. Metaphorically and factually, *Homo sapiens* is one of the five surviving species of great apes, along with chimpanzees, bonobos, gorillas, and orangutans (gibbons are considered a "lesser ape"). We shared a common ancestor with two of these apes—bonobos and chimps—just five million years ago.[1] That's "the day before yesterday" in evolutionary terms. The fine print distinguishing humans from the other great apes is regarded as "wholly artificial" by most primatologists these days.[2]

If we're "above" nature, it's only in the sense that a shaky-legged surfer is "above" the ocean. Even if we never slip (and we all do), our inner nature can pull us under at any moment. Those of us raised in the West have been assured that we humans are special, unique among living things, above and beyond the world around us, exempt from the humilities and humiliations that pervade and define animal life. The natural world lies below and beneath us, a cause for shame, disgust, or alarm; something smelly and messy to be hidden behind closed doors, drawn curtains, and minty freshness. Or we overcompensate and imagine nature floating angelically in soft focus up above, innocent, noble, balanced, and wise.

Like bonobos and chimps, we are the randy descendents of hypersexual ancestors. At first blush, this may seem an overstatement, but it's a truth that should have become common knowledge long ago. Conventional notions of monogamous, till-death-do-us-part marriage strain under the dead weight of a false narrative that insists we're something else. What is the essence of human sexuality and how did it get to be that way? In the following pages, we'll explain how seismic cultural shifts that began about ten thousand years ago rendered the true story of human sexuality so subversive and threatening that for centuries it has been silenced by religious authorities, pathologized by physicians, studiously ignored by scientists, and covered up by moralizing therapists.

Deep conflicts rage at the heart of modern sexuality. Our cultivated ignorance is devastating. The campaign to obscure the true nature of our species' sexuality leaves half our marriages collapsing under an unstoppable tide of swirling sexual frustration, libido-killing boredom, impulsive betrayal, dysfunction, confusion, and shame. Serial monogamy stretches before (and behind) many of us like an archipelago of failure: isolated islands of transitory happiness in a cold, dark sea of disappointment. And how many of the couples who manage to stay together for the long haul have done so by resigning themselves to sacrificing their eroticism on the altar of three of life's irreplaceable joys: family stability, companionship, and emotional, if not sexual, intimacy? Are those who innocently aspire to these joys cursed by nature to preside over the slow strangulation of their partner's libido?

The Spanish word *esposas* means both "wives" and "handcuffs." In English, some men ruefully joke about the *ball and chain*. There's good reason marriage is often depicted and mourned as the beginning of the end of a man's sexual life. And women fare no better. Who wants to share her life with a man who feels trapped and diminished by his love for her, whose honor marks the limits of his freedom? Who wants to spend her life apologizing for being just one woman?

Yes, something is seriously wrong. The American Medical Association reports that some 42 percent of American women suffer from sexual dysfunction, while Viagra breaks sales records year after year.

Worldwide, pornography is reported to rake in anywhere from fifty-seven *billion* to a hundred *billion* dollars annually. In the United States, it generates more revenue than CBS, NBC, and ABC combined and more than all professional football, baseball, and basketball franchises. According to *U.S. News and World Report,* "Americans spend more money at strip clubs than at Broadway, off-Broadway, regional and nonprofit theaters, the opera, the ballet and jazz and classical music performances—combined."[3]

There's no denying that we're a species with a sweet tooth for sex. Meanwhile, so-called traditional marriage appears to be under assault from all sides—as it collapses from within. Even the most ardent defenders of *normal* sexuality buckle under its weight, as never-ending bipartisan perp-walks of politicians (Clinton, Vitter, Gingrich, Craig, Foley, Spitzer, Sanford) and religious figures (Haggard, Swaggert, Bakker) trumpet their support of *family values* before slinking off to private assignations with lovers, prostitutes, and interns.

Denial hasn't worked. Hundreds of Catholic priests have confessed to thousands of sex crimes against children in the past few decades alone. In 2008, the Catholic Church paid $436 million in compensation for sexual abuse. More than a fifth of the victims were under ten years old. This we know. Dare we even imagine the suffering such crimes have caused in the seventeen centuries since a sexual life was perversely forbidden to priests in the earliest known papal decree: the *Decreta* and *Cum in unum* of Pope Siricius (c. 385)? What is the moral debt owed to the forgotten victims of this misguided rejection of basic human sexuality?

On threat of torture, in 1633, the Inquisition of the Roman Catholic Church forced Galileo to state publicly what he knew to be false: that the Earth sat immobile at the center of the universe. Three and a half centuries later, in 1992, Pope John Paul II admitted that the scientist had been right all along, but that the Inquisition had been "well-intentioned."

Well, there's no Inquisition like a *well-intentioned* Inquisition!

Like those childishly intransigent visions of an entire universe spinning around an all-important Earth, the standard narrative of prehistory offers an immediate, primitive sort of comfort. Just as pope after pope dismissed any cosmology that removed humankind from the exalted center of the endless expanse of space, just as Darwin was (and, in some crowds, still is) ridiculed for recognizing that human beings are the creation of natural laws, many scientists are blinded by their emotional resistance to any account of human sexual evolution that doesn't revolve around the monogamous nuclear family unit.

Although we're led to believe we live in times of sexual liberation, contemporary human sexuality throbs with obvious, painful truths that must not be spoken aloud. The conflict between what we're *told* we feel and what we *really* feel may be the richest source of confusion, dissatisfaction, and unnecessary suffering of our time. The answers normally proffered don't answer the questions at the heart of our erotic lives: why are men and women so different in our desires, fantasies, responses, and sexual behavior? Why are we betraying and divorcing each other at ever increasing rates when not opting out of marriage entirely? Why the pandemic spread of single-parent families? Why does the passion evaporate from so many marriages so quickly? What causes the death of desire? Having evolved together right here on Earth, why do so many men and women resonate with the idea that we may as well be from different planets?

Oriented toward medicine and business, American society has responded to this ongoing crisis by developing a marital-industrial complex of couples therapy, pharmaceutical hard-ons, sex-advice columnists, creepy father-daughter purity cults, and an endless stream of in-box come-ons ("Unleash your LoveMonster! She'll thank you!"). Every month, truckloads of glossy supermarket magazines offer the same old tricks to get the spark back into our moribund sex lives.

Yes, a few candles here, some crotchless panties there, toss a handful of rose petals on the bed and it'll be just like the very first time! What's that you say? He's still checking out other women? She's still got an air of detached disappointment? He's finished before you've begun?

Well, then, let the experts figure out what ails you, your partner, your relationship. Perhaps his penis needs enlarging or her vagina needs a retrofit. Maybe he has "commitment issues," a "fragmentary superego," or the dreaded "Peter Pan complex." Are you depressed? You say you love your spouse of a dozen years but don't feel sexually attracted the way you used to? One or both of you are tempted by another? Maybe you two should try doing it on the kitchen floor. Or force yourself to do it every night for a year.[4] Maybe he's going through a midlife crisis. Take these pills. Get a new hairstyle. *Something* must be wrong with you.

Ever feel like the victim of a well-intentioned Inquisition?

This split-personality relationship with our true sexual nature is anything but news to entertainment corporations, who have long reflected the same fractured sensibility between public pronouncement and private desire. In 2000, under the headline "Wall Street Meets Pornography," *The New York Times* reported that General Motors sold more graphic sex films than Larry Flynt, owner of the Hustler empire. Over eight million American subscribers to DirecTV, a General Motors subsidiary, were spending about $200 million a year on pay-per-view sex films from satellite providers. Similarly, Rupert Murdoch, owner of the Fox News Network and the nation's leading conservative newspaper, *The Wall Street Journal,* was pulling in more porn money through a satellite company than Playboy made with its magazine, cable, and Internet businesses combined.[5] AT&T, also a supporter of conservative values, sells hard-core porn to over a million hotel rooms throughout the country via its Hot Network.

The frantic sexual hypocrisy in America is inexplicable if we adhere to traditional models of human sexuality insisting that monogamy is natural, marriage is a human universal, and any family structure other than the nuclear is aberrant. We need a new understanding of ourselves, based not on pulpit proclamations or feel-good Hollywood fantasies, but on a bold and unashamed assessment of the plentiful scientific data that illuminate the true origins and nature of human sexuality.

We are at war with our eroticism. We battle our hungers, expectations, and disappointments. Religion, politics, and even science square off against biology and millions of years of evolved appetites. How to defuse this intractable struggle?

In the following pages, we reassess some of the most important science of our time. We question the deepest assumptions brought to contemporary views of marriage, family structure, and sexuality—issues affecting each of us every day and every night.

We'll show that human beings evolved in intimate groups where almost everything was shared—food, shelter, protection, child care, even sexual pleasure. We don't argue that humans are natural-born Marxist hippies. Nor do we hold that romantic love was unknown or unimportant in prehistoric communities. But we'll demonstrate that contemporary culture misrepresents the link between love and sex. With and without love, a casual sexuality was the norm for our prehistoric ancestors.

Let's address the question you're probably already asking: how can we possibly know anything about sex in prehistory? Nobody alive today was there to witness prehistoric life, and since social behavior leaves no fossils, isn't this all just wild speculation?

Not quite. There's an old story about the trial of a man charged with biting off another man's finger in a fight. An eyewitness took the stand. The defense attorney asked, "Did you actually see my client bite off the finger?" The witness said, "Well, no, I didn't." "Aha!" said the attorney with a smug smile. "How then can you claim he bit off the man's finger?" "Well," replied the witness, "I saw him spit it out."

In addition to a great deal of circumstantial evidence from societies around the world and closely related nonhuman primates, we'll take a look at some of what evolution has spit out. We'll examine the anatomical evidence still evident in our bodies and the yearning for sexual novelty expressed in our pornography, advertising, and after-work happy hours. We'll even decode messages in the so-called "copulatory vocalizations" of thy neighbor's wife as she calls out ecstatically in the still of night.

Readers acquainted with the recent literature on human sexuality will be familiar with what we call the standard narrative of human sexual evolution (hereafter shortened to "the standard narrative"). It goes something like this:

1. Boy meets girl.
2. Boy and girl assess one another's *mate value* from perspectives based upon their differing reproductive agendas/capacities:
    - He looks for signs of youth, fertility, health, absence of previous sexual experience, and likelihood of future sexual fidelity. In other words, his assessment is skewed toward finding a fertile, healthy young mate with many childbearing years ahead and no current children to drain his resources.
    - She looks for signs of wealth (or at least prospects of future wealth), social status, physical health, and likelihood that he will stick around to protect and provide for their children. Her guy must be willing and able to provide materially for her (especially during pregnancy and breastfeeding) and their children (known as *male parental investment*).
3. Boy gets girl: assuming they meet one another's criteria, they "mate," forming a long-term pair bond—the "fundamental condition of the human species," as famed author Desmond Morris put it. Once the pair bond is formed:
    - She will be sensitive to indications that he is considering leaving (vigilant toward signs of infidelity involving *intimacy* with other women that would threaten her access to his resources and protection)—while keeping an eye out (around

ovulation, especially) for a quick fling with a man genetically superior to her husband.

- He will be sensitive to signs of her *sexual* infidelities (which would reduce his all-important paternity certainty)—while taking advantage of short-term sexual opportunities with other women (as his sperm are easily produced and plentiful).

Researchers claim to have confirmed these basic patterns in studies conducted around the world over several decades. Their results seem to support the standard narrative of human sexual evolution, which appears to make a lot of sense. But they don't, and it doesn't.

While we don't dispute that these patterns play out in many parts of the modern world, we don't see them as elements of human nature so much as adaptations to social conditions—many of which were introduced with the advent of agriculture no more than ten thousand years ago. These behaviors and predilections are not biologically programmed traits of our species; they are evidence of the human brain's flexibility and the creative potential of community.

To take just one example, we argue that women's seemingly consistent preference for men with access to wealth is *not* a result of innate evolutionary programming, as the standard model asserts, but simply a behavioral adaptation to a world in which men control a disproportionate share of the world's resources. As we'll explore in detail, before the advent of agriculture a hundred centuries ago, women typically had as much access to food, protection, and social support as did men. We'll see that upheavals in human societies resulting from the shift to settled living in agricultural communities brought radical changes to women's ability to survive. Suddenly, women lived in a world where they had to barter their reproductive capacity for access to the resources and protection they needed to survive. But these conditions are very different from those in which our species had been evolving previously.

It's important to keep in mind that when viewed against the full scale of our species' existence, ten thousand years is but a brief moment. Even if we ignore the roughly two million years since the emergence of our *Homo* lineage, in which our direct ancestors lived in small foraging social groups, anatomically modern humans are estimated to have existed as long as 200,000 years.* With the earliest evidence of agriculture dating to about 8000 BCE, the amount of time our species has spent living in settled agricultural societies represents just 5 percent of our collective experience, at most. As recently as a few hundred years ago, most of the planet was still occupied by foragers.

So in order to trace the deepest roots of human sexuality, it's vital to look beneath the thin crust of recent human history. Until agriculture, human beings evolved in societies organized around an insistence on sharing just about everything. But all this sharing doesn't make anyone a *noble savage*. These pre-agricultural societies were no nobler than you are when you pay your taxes or insurance premiums. Universal, culturally imposed sharing was simply the most effective way for our highly social species to minimize risk. Sharing and self-interest, as we shall see, are not mutually exclusive. Indeed, what many anthropologists call *fierce egalitarianism* was the predominant pattern of social organization around the world for many millennia before the advent of agriculture.

But human societies changed in radical ways once they started farming and raising domesticated animals. They organized themselves around hierarchical political structures, private property, densely populated settlements, a radical shift in the status of women, and other social configurations that together represent an enigmatic disaster for our species: human population growth mushroomed as quality of life plummeted. The shift to agriculture, wrote author Jared Diamond, is a "catastrophe from which we have never recovered."[6]

Several types of evidence suggest our pre-agricultural (prehistoric) ancestors lived in groups where most mature individuals would have

---

* We use the terms "foragers" and "hunter-gatherers" interchangeably throughout the text.

had several ongoing sexual relationships at any given time. Though often casual, these relationships were not random or meaningless. Quite the opposite: they reinforced crucial social ties holding these highly interdependent communities together.[7]

We've found overwhelming evidence of a decidedly casual, friendly prehistory of human sexuality echoed in our own bodies, in the habits of remaining societies still lingering in relative isolation, and in some surprising corners of contemporary Western culture. We'll show how our bedroom behavior, porn preferences, fantasies, dreams, and sexual responses all support this reconfigured understanding of our sexual origins. Questions you'll find answered in the following pages include:

- Why is long-term sexual fidelity so difficult for so many couples?
- Why does sexual passion often fade, even as love deepens?
- Why are women potentially multi-orgasmic, while men all too often reach orgasm frustratingly quickly and then lose interest?
- Is sexual jealousy an unavoidable, uncontrollable part of human nature?
- Why are human testicles so much larger than those of gorillas but smaller than those of chimps?
- Can sexual frustration make us sick? How did a lack of orgasms cause one of the most common diseases in history, and how was it treated?

A Few Million Years in a Few Pages

In a nutshell, here's the story we tell in the following pages: A few million years ago, our ancient ancestors (*Homo erectus*) shifted from a gorilla-like mating system where an alpha male fought to win and maintain a harem of females to one in which most males had sexual access to females. Few, if any experts dispute the fossil evidence for this shift.[8]

But we part company from those who support the standard narrative when we look at what this shift signifies. The standard narrative holds that this is when long-term pair bonding began in our species: if each male could have only one female mate at a time, most males would end up with a girl to call their own. Indeed, where there is debate about the nature of innate human sexuality, the only two *acceptable* options appear to be that humans evolved to be either monogamous (M–F) or polygynous (M–FFF+)—with the conclusion normally being that women generally prefer the former configuration while most men would opt for the latter.

But what about multiple mating, where most males and females have more than one concurrent sexual relationship? Why—apart from moral disgust—is prehistoric promiscuity not even considered, when nearly every relevant source of evidence points in that direction?

After all, we know that the foraging societies in which human beings evolved were small-scale, highly egalitarian groups who shared almost everything. There is a remarkable consistency to how *immediate return* foragers live—wherever they are.* The !Kung San of Botswana have a great deal in common with Aboriginal people living in outback Australia and tribes in remote pockets of the Amazon rainforest. Anthropologists have demonstrated time and again that immediate-return

* Anthropologist James Woodburn (1981/1998) classified foraging societies into immediate-return (simple) and delayed-return (complex) systems. In the former, food is eaten within days of acquisition, without elaborate processing or storage. Unless otherwise noted, we refer to these societies.

hunter-gatherer societies are nearly universal in their *fierce egalitarianism*. Sharing is not just encouraged; it's mandatory. Hoarding or hiding food, for example, is considered deeply shameful, almost unforgivable behavior in these societies.[9]

Foragers divide and distribute meat equitably, breastfeed one another's babies, have little or no privacy from one another, and depend upon each other for survival. As much as our social world revolves around notions of private property and individual responsibility, theirs spins in the opposite direction, toward group welfare, group identity, profound interrelation, and mutual dependence.

Though this may sound like naïve New Age idealism, whining over the lost Age of Aquarius, or a celebration of prehistoric communism, not one of these features of pre-agricultural societies is disputed by serious scholars. The overwhelming consensus is that egalitarian social organization is the de-facto system for foraging societies in all environments. In fact, no other system *could* work for foraging societies. Compulsory sharing is simply the best way to distribute risk to everyone's benefit: participation mandatory. Pragmatic? Yes. Noble? Hardly.

We believe this sharing behavior extended to sex as well. A great deal of research from primatology, anthropology, anatomy, and psychology points to the same fundamental conclusion: human beings and our hominid ancestors have spent almost all of the past few million years or so in small, intimate bands in which most adults had several sexual relationships at any given time. This approach to sexuality probably persisted until the rise of agriculture and private property no more than ten thousand years ago. In addition to voluminous scientific evidence, many explorers, missionaries, and anthropologists support this view, having penned accounts rich with tales of orgiastic rituals, unflinching mate sharing, and an open sexuality unencumbered by guilt or shame.

If you spend time with the primates closest to human beings, you'll see female chimps having intercourse dozens of times per day, with most or all of the willing males, and rampant bonobo group sex that leaves everyone relaxed and maintains intricate social networks.

Explore contemporary human beings' lust for particular kinds of pornography or our notorious difficulties with long-term sexual monogamy and you'll soon stumble over relics of our hypersexual ancestors.

Our bodies echo the same story. The human male has testicles far larger than any monogamous primate would ever need, hanging vulnerably outside the body where cooler temperatures help preserve stand-by sperm cells for multiple ejaculations. He also sports the longest, thickest penis found on any primate on the planet, as well as an embarrassing tendency to reach orgasm too quickly. Women's pendulous breasts (utterly unnecessary for breastfeeding children), impossible-to-ignore cries of delight (*female copulatory vocalization* to the clipboard-carrying crowd), and capacity for orgasm after orgasm all support this vision of prehistoric promiscuity. Each of these points is a major snag in the standard narrative.

Once people were farming the same land season after season, private property quickly replaced communal ownership as the modus

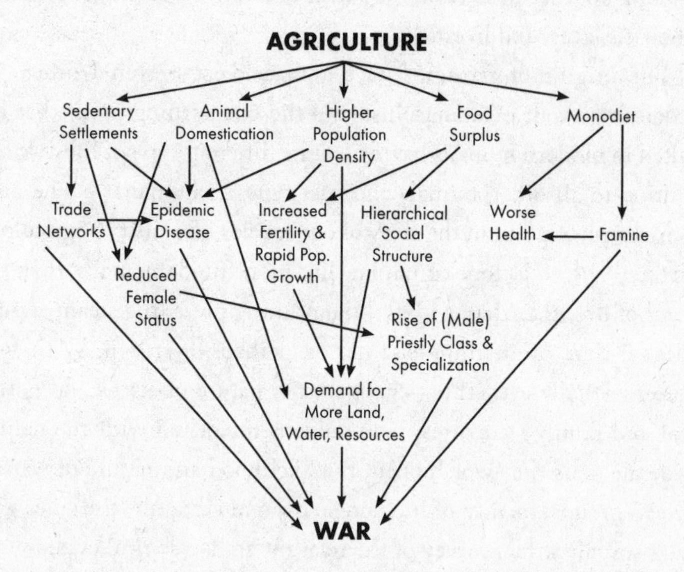

operandi in most societies. For nomadic foragers, personal property—anything needing to be carried—is kept to a minimum, for obvious reasons. There is little thought given to who owns the land, or the fish in the river, or the clouds in the sky. Men (and often, women) confront danger together. An individual male's *parental investment*, in other words—the core element of the standard narrative—tends to be diffuse in societies like those in which we evolved, not directed toward one particular woman and her children, as the conventional model insists.

But when people began living in settled agricultural communities, social reality shifted deeply and irrevocably. Suddenly it became crucially important to know where your field ended and your neighbor's began. Remember the Tenth Commandment: "Thou shalt not covet thy neighbour's house, thou shalt not covet thy neighbour's wife, nor his manservant, nor his maidservant, nor his ox, nor his ass, nor any thing that [is] thy neighbour's." Clearly, the biggest loser (aside from slaves, perhaps) in the agricultural revolution was the human female, who went from occupying a central, respected role in foraging societies to becoming another possession for a man to earn and defend, along with his house, slaves, and livestock.

"The origins of farming," says archaeologist Steven Mithen, "is the defining event of human history—the one turning point that has resulted in modern humans having a quite different type of lifestyle and cognition to all other animals and past types of humans."[10] The most important pivot point in the story of our species, the shift to agriculture redirected the trajectory of human life more fundamentally than the control of fire, the Magna Carta, the printing press, the steam engine, nuclear fission, or anything else has or, perhaps, ever will. With agriculture, virtually everything changed: the nature of status and power, social and family structures, how humans interacted with the natural world, the gods they worshipped, the likelihood and nature of warfare between groups, quality of life, longevity, and certainly, the rules governing sexuality. His survey of the relevant archaeological evidence led archaeologist Timothy Taylor, author of *The Prehistory of Sex*, to state, "While hunter-gatherer sex had been modeled on an idea of sharing and complementarity, early agriculturalist sex was voyeuristic, repressive, homophobic, and focused on reproduction." "Afraid of the wild," he concludes, "farmers set out to destroy it."[11]

Land could now be possessed, owned, and passed down the generations. Food that had been hunted and gathered now had to be sowed, tended, harvested, stored, defended, bought, and sold. Fences, walls, and irrigation systems had to be built and reinforced; armies to defend it all had to be raised, fed, and controlled. Because of private property, *for the first time in the history of our species, paternity became a crucial concern.*

But the standard narrative insists that paternity certainty has *always* been of utmost importance to our species, that our very genes dictate we organize our sexual lives around it. Why, then, is the anthropological record so rich with examples of societies where biological paternity is of little or no importance? Where paternity is unimportant, men tend to be relatively unconcerned about women's sexual fidelity.

But before we get into these real-life examples, let's take a quick trip to the Yucatán.

PART I

On the Origin of the Specious

CHAPTER ONE

Remember the Yucatán!

The function of the imagination is not to make strange things settled, so much as to make settled things strange.

G. K. Chesterton

Forget the Alamo. The Yucatán provides a more useful lesson.

It was early spring, 1519. Hernán Cortés and his men had just arrived off the coast of the Mexican mainland. The conquistador ordered his men to bring one of the natives to the deck of the ship, where Cortés asked him the name of this exotic place they'd found. The man responded, "*Ma c'ubah than*," which the Spanish heard as *Yucatán*. Close enough. Cortés proclaimed that from that day onward, Yucatán and any gold it contained belonged to Spain, and so on.

Four and a half centuries later, in the 1970s, linguists researching archaic Mayan dialects concluded that *Ma c'ubah than* meant "I do not understand you."[1]

Each spring, thousands of American university students celebrate with wet T-shirt contests, foam parties, and Jell-O wrestling on the beautiful beaches of the I Do Not Understand You Peninsula.

But confusion mistaken for knowledge isn't limited to spring break. We all fall into this trap. (One night, over dinner, a close friend mentioned that her favorite Beatles song is "Hey Dude.") Despite their years of training, even scientific types slip into thinking they are observing something when in fact they are simply projecting their biases and ignorance. What trips up the scientists is the same cognitive failing we all share: it's hard to be certain about what we think we know, but don't really. Having misread the map, we're sure we know where we are. In the face of evidence to the contrary, most of us tend to go with our gut, but the gut can be an unreliable guide.

You Are What You Eat

Take food, for example. We all assume that our craving or disgust is due to something about the food itself—as opposed to being an often arbitrary response preprogrammed by our culture. We understand that Australians prefer cricket to baseball, or that the French somehow find Gérard Depardieu sexy, but how hungry would you have to be before you would consider plucking a moth from the night air and popping it, frantic and dusty, into your mouth? *Flap, crunch, ooze.* You could wash it down with some saliva beer. How does a plate of sheep's brain sound? Broiled puppy with gravy? May we interest you in pig's ears or shrimp heads? Perhaps a deep-fried songbird that you chew up, bones, beak, and all? A game of cricket on a field of grass is one thing, but pan-fried crickets over lemongrass? That's revolting.

Or is it? If lamb chops are fine, what makes lamb brains horrible? A pig's shoulder, haunch, and belly are damn fine eatin', but the ears, snout, and feet are gross? How is lobster so different from grasshopper? Who distinguishes delectable from disgusting, and what's their rationale? And what about all the exceptions? Grind up those leftover pig parts, stuff 'em in an intestine, and you've got yourself respectable sausages or hot dogs. You may think bacon and eggs just go together, like French fries and ketchup or salt and pepper. But the combination of bacon and eggs for breakfast was dreamed up about a hundred years ago by an advertising agency hired to sell more bacon, and the Dutch eat their fries with mayonnaise, not ketchup.

Think it's rational to be grossed out by eating bugs? Think again. A hundred grams of dehydrated cricket contains 1,550 milligrams of iron, 340 milligrams of calcium, and 25 milligrams of zinc—three minerals often missing in the diets of the chronic poor. Insects are richer in minerals and healthy fats than beef or pork. Freaked out by the exoskeleton, antennae, and way too many legs? Then stick to the Turf and forget the Surf because shrimp, crabs, and lobsters are all arthropods, just like grasshoppers. And they eat the nastiest of what sinks to the bottom of the ocean, so don't talk about bugs' disgusting diets. Anyway, you may have bug parts stuck between your teeth right now. The Food and Drug Administration tells its inspectors to ignore insect parts in black pepper unless they find more than 475 of them per 50 grams, on average.[2] Some estimates suggest that Americans unknowingly eat anywhere from *one to two pounds* of insects per year.

An Italian professor recently published *Ecological Implications of Minilivestock: Potential of Insects, Rodents, Frogs and Snails.* (Minicowpokes sold separately.) Writing in Slate.com, William Saletan tells us about a company by the name of Sunrise Land Shrimp. The company's logo: "Mmm. That's good Land Shrimp!" Three guesses what Land Shrimp is.

> *Witchetty grub tastes like nut-flavored scrambled eggs and mild mozzarella, wrapped in a phyllo dough pastry . . . This is capital-D Delicious.*
>
> PETER MENZEL AND FAITH D'ALUISIO,
> *Man Eating Bugs*

Early British travelers to Australia reported that the Aborigines they met lived miserably and suffered from chronic famine. But the native people, like most hunter-gatherers, were uninterested in farming. The same Europeans reporting the widespread starvation in their letters and journals were perplexed that the natives didn't seem emaciated. In fact, they struck the visitors as being rather fat and lazy. Yet, the Europeans were convinced the Aborigines were starving to death. Why? Because they saw the native people resorting to last resorts—eating insects, Witchetty grubs, and rats, critters that surely nobody would eat who wasn't starving. That this diet was nutritious, plentiful, and could taste like "nut-flavored scrambled eggs and mild mozzarella" never occurred to the British, who were no doubt homesick for haggis and clotted cream.

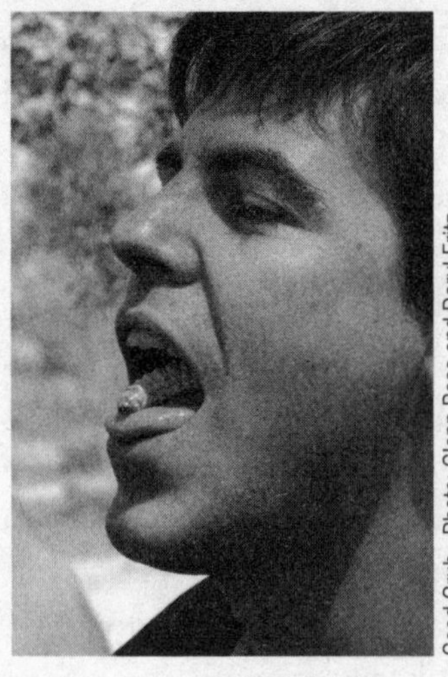

Good Grub. Photo: Glenn Rose and Daryl Fritz

Our point? That something *feels* natural or unnatural doesn't mean it is. Every one of the examples above, including saliva beer, is savored somewhere—by folks who would be disgusted by much of what you eat regularly. Especially when we're talking about intimate, personal, biological experiences like eating or having sex, we mustn't forget that the familiar fingers of culture reach deep into our minds. We can't feel them adjusting our dials and flicking our switches, but every culture leads its members to believe some things are naturally right and others naturally wrong. These beliefs may feel right, but it's a feeling we trust at our own peril.

Like those early Europeans, each of us is constrained by our own sense of what is normal and natural. We're all members of one tribe or another—bonded by culture, family, religion, class, education, employment, team affiliation, or any number of other criteria. An essential first step in discerning the *cultural* from the *human* is what mythologist Joseph Campbell called *detribalization*. We have to recognize the various tribes we belong to and begin extricating ourselves from the unexamined assumptions each of them mistakes for *the truth*.

Authorities assure us that we are jealous of our mates because such feelings are only *natural*. Experts opine that women need commitment to feel sexual intimacy because "that's just the way they are." Some of the most prominent evolutionary psychologists insist that science has confirmed that we are, at base, a jealous, possessive, murderous, and deceitful species just barely saved by our precarious capacity to rise above our dark essence and submit to civilized propriety. To be sure, we humans have hankerings and aversions deeper than cultural influence, at the core of our animal being. We don't argue that humans are born "blank slates," awaiting operating instructions. But how something "feels" is far from a reliable guide to distinguishing biological truth from cultural influence.

Go looking for a book about human nature and you'll be confronted by *Demonic Males*, *Mean Genes*, *Sick Societies*, *War Before Civilization*, *Constant Battles*, *The Dark Side of Man*, and *The Murderer Next Door*. You'll be lucky to escape alive! But do these blood-splattered volumes offer a realistic depiction of scientific truth, or a projection of contemporary assumptions and fears onto the distant past?

In the following chapters, we reconsider these and other aspects of social behavior, rearranging them to form a different view of our past. We believe our model goes much farther toward explaining how we got to where we are today and most importantly, *why many, if not most, sexually dysfunctional marriages are nobody's fault*. We'll show why a great deal of the information we receive about human sexuality—particularly that received from some evolutionary psychologists—is mistaken, based upon unfounded, outdated assumptions going back to Darwin and beyond. Too many scientists are hard at work trying to complete the wrong puzzle, struggling to force their findings into preconceived, culturally approved notions of what they think human sexuality *should be* rather than letting the pieces of information fall where they may.

Our model might strike you as absurd, salacious, insulting, scandalous, fascinating, depressing, illuminating, or obvious. But whether or not you are comfortable with what we present here, we hope you'll keep reading. We are not advocating any particular response to the information we've put together. Frankly, we're not sure what to do with it ourselves.

Undoubtedly, some readers will react emotionally to our "scandalous" model of human sexuality. Our interpretation of the data will be dismissed and derided by stalwart souls defending the ramparts of the standard narrative. They'll be shouting, "Remember the Alamo!" But our advice, as we lead you through this story of unwarranted assumptions, desperate conjecture, and mistaken conclusions, is to forget the Alamo, but always remember the Yucatán.

CHAPTER TWO

What Darwin Didn't Know About Sex

We are not here concerned with hopes or fears, only with truth as far as our reason permits us to discover it.

CHARLES DARWIN, *The Descent of Man*

A fig leaf can hide many things, but a human erection isn't one of them. The standard narrative of the origins and nature of human sexuality claims to explain the development of a deceitful, reluctant sort of sexual monogamy. According to this oft-told tale, heterosexual men and women are pawns in a proxy war directed by our opposed genetic agendas. The whole catastrophe, we're told, results from the basic biological designs of males and females.* Men strain to spread their cheap and plentiful seed far and wide (while still trying to control one or a few females in order to increase their paternity certainty). Meanwhile, women are guarding their limited supply of metabolically expensive eggs from unworthy suitors. But once they've roped in a provider-husband, they're quick to hike up their skirts (when ovulating) for quick-and-dirty clandestine mating opportunities with square-jawed men of obvious genetic superiority. It's not a pretty picture.

* Our use of the word "design" is purely metaphorical—not meant to imply any "designer" or intentionality underlying evolved human behavior or anatomy.

Biologist Joan Roughgarden points out that it's an image little changed from that described by Darwin 150 years ago. "The Darwinian narrative of sex roles is not some quaint anachronism," she writes. "Restated in today's biological jargon, the narrative is considered proven scientific fact. . . . Sexual selection's view of nature emphasizes conflict, deceit, and dirty gene pools."[1]

No less an authority than The Advice Goddess herself (syndicated columnist Amy Alkon) voices the popularized expression of this oft-told tale: "There are a lot of really bad places to be a single mother, but probably one of the worst ever was 1.8 million years ago on the savannah. The ancestral women who successfully passed their genes on to us were those who were choosy about who they went under a bush with, weeding out the dads from the cads. Men had a different genetic imperative—to avoid bringing home the bison for kids who weren't theirs—and evolved to regard girls who give it up too easily as too high risk for anything beyond a roll on the rock pile."[2] Note how so much fits into this tidy package: the vulnerabilities of motherhood, separating dads from cads, paternal investment, jealousy, and the sexual double standard. But as they say at the airport, beware of tidy packages you didn't pack yourself.

As for an English lady, I have almost forgotten what she is. — something very angelic and good.

CHARLES DARWIN, in a letter from the HMS *Beagle*

Gentry had to be pitied. They had so few advantages in respect of love. They could say they longed for a kiss from a bouncy wife in a vicarage garden. They couldn't say she roared under me and clutched my back, and I shot my specimen to blazes.

ROGER MCDONALD, *Mr. Darwin's Shooter*

The best place to begin a reassessment of our conflicted relationship with sexuality may be with Charles Darwin himself. Darwin's brilliant work inadvertently lent an enduring scientific patina to what is essentially anti-erotic bias. Despite his genius, what Darwin didn't know about sex could fill volumes. This is one of them.

On the Origin of Species was published in 1859, a time when little was known about human life before the classical era. Prehistory, the period we define as the 200,000 or so years when anatomically modern people lived without agriculture and writing, was a blank slate theorists could fill only with conjecture. Until Darwin and others began to loosen the link between religious doctrine and scientific truth, guesses about the distant past were restricted by church teachings. The study of primates was in its infancy. Given the scientific data Darwin never saw, it's not surprising that this great thinker's blind spots can be as illuminating as his insights.[3]

For example, Darwin's ready acceptance of Thomas Hobbes's still-famous characterization of prehistoric human life as having been "solitary, poor, nasty, brutish, and short" left these mistaken assumptions embedded in present-day theories of human sexuality. Asked to imagine prehistoric human sex, most of us conjure the hackneyed image of the caveman dragging a dazed woman by her hair with one hand, a club in the other. As we'll see, this image of prehistoric human life is mistaken in every one of its Hobbesian details. Similarly, Darwin incorporated Thomas Malthus's unsubstantiated theories about the distant past into his own theorizing, leading him to dramatic overestimations of early human suffering (and thus, of the comparative superiority of Victorian life). These pivotal misunderstandings persist in many contemporary evolutionary scenarios.

Though he certainly didn't originate this narrative of the interminable tango between randy male and choosy female, Darwin beat the drum for its supposed "naturalness" and inevitability. He wrote passages like, "The female . . . with the rarest exception, is less eager than the male . . . [She] requires to be courted; she is coy, and may often be seen endeavoring for a long time to escape the male." While this female reticence is a key feature in the mating systems of many mammals, it isn't particularly applicable to human beings or, for that matter, the primates most closely related to us.

In light of the philandering he saw going on around him, Darwin wondered whether early humans might have been polygynists (one male mating with several females), writing, "*Judging from the social habits of man as he now exists,* and from most savages being polygamists, the most probable view is that primeval man aboriginally lived in small communities, each with as many wives as he could support and obtain, whom he would have jealously guarded against all other men [emphasis added]."[4]

Evolutionary psychologist Steven Pinker appears to be "judging from the social habits of man as he now exists" as well (though without Darwin's self-awareness) when he bluntly asserts, "In all societies, sex is at least somewhat 'dirty.' It is conducted in private, pondered obsessively, regulated by custom and taboo, the subject of gossip and teasing, and a trigger for jealous rage."[5] We'll show that while sex is indeed "regulated by custom and taboo," there are multiple exceptions to every other element of Pinker's overconfident declaration.

Like all of us, Darwin incorporated his own personal experience—or its absence—into his assumptions about the nature of all human life. In *The French Lieutenant's Woman*, John Fowles gives a sense of the sexual hypocrisy that characterized Darwin's world. Nineteenth-century England, writes Fowles, was "an age where woman was sacred; and where you could buy a thirteen-year-old girl for a few pounds—a few shillings, if you wanted her for only an hour or two. . . . Where the female body had never been so hidden from view; and where every sculptor was judged by his ability to carve naked women. . . . Where it was universally maintained that women do not have orgasms; and yet every prostitute was taught to simulate them."[6]

In some respects, the sexual mores of Victorian Britain replicated the mechanics of the age-defining steam engine. Blocking the flow of erotic energy creates ever-increasing pressure which is put to work through short, controlled bursts of productivity. Though he was wrong about a lot, it appears Sigmund Freud got it right when he observed that "civilization" is built largely on erotic energy that has been blocked, concentrated, accumulated, and redirected.

"To keep body and mind untainted," explains Walter Houghton in *The Victorian Frame of Mind*, "the boy was taught to view women as objects of the greatest respect and even awe. He was to consider nice women (his sister and mother, his future bride) as creatures more like angels than human beings—an image wonderfully calculated not only to dissociate love from sex, but to turn love into worship, and worship of purity."[7] When not in the mood to worship the purity of his sisters, mother, daughters, and wife, men were expected to purge their lust with prostitutes, rather than threatening familial and social stability by "cheating" with "decent women." Nineteenth-century philosopher Arthur Schopenhauer observed that "there are 80,000 prostitutes in London alone; and what are they if not sacrifices on the altar of monogamy?"[8]

Charles Darwin was certainly not unaffected by the erotophobia of his era. In fact, one could argue that he was *especially* sensitive to its influence, inasmuch as he came of age in the intellectual shadow of his famous—and shameless—grandfather, Erasmus Darwin, who had flouted the sexual mores of his day by openly having children with various women and even going so far as to celebrate group sex in his poetry.[9] The death of Charles's mother when he was just eight years old may well have enhanced his sense of women as angelic creatures floating above earthly urges and appetites.

Psychiatrist John Bowlby, one of Darwin's most highly regarded biographers, attributes Darwin's lifelong anxiety attacks, depression, chronic headaches, dizziness, nausea, vomiting, and hysterical crying fits to the separation anxiety created by the early loss of his mother. This interpretation is supported by a strange letter the adult Charles wrote to a cousin whose wife had just died: "Never in my life having lost one near relation," he wrote, apparently repressing his memories of his own mother's death, "I daresay I cannot imagine how severe grief such as yours must be." Another indication of this psychological scarring was recalled by his granddaughter, who remembered how confused Charles had been when someone added the letter "M" to the beginning of the word OTHER in a game similar to Scrabble. Charles looked at the board for a long time before declaring, to everyone's confusion, that no such word existed.[10]

A hyper-Victorian aversion to (and obsession with) the erotic seems to have continued in Charles's eldest surviving daughter, Henrietta. "Etty," as she was known, edited her father's books, taking her blue crayon to passages she considered inappropriate. In Charles's biography of his free-thinking grandfather, for example, she deleted a reference to Erasmus's "ardent love of women." She also removed "offensive" passages from *The Descent of Man* and Darwin's autobiography.

Etty's prim enthusiasm for stamping out anything sexual wasn't limited to the written word. She waged a bizarre little war against the so-called stinkhorn mushroom (*Phallus ravenelii*) that still pops up in the woods around the Darwin estate. Apparently, the similarity of the mushroom to the human penis was a bit much for poor Etty. As her niece (Charles's granddaughter) recalled years later, "Aunt Etty . . . armed with a basket and a pointed stick, and wearing a special hunting cloak and gloves," would set out in search of the mushrooms. At the end of the day, Aunt Etty "burn[ed them] in the deepest secrecy on the drawing room fire with the door locked—because of the morals of the maids."[11]

He will hold thee, when his passion shall have spent its novel force, something better than his dog, a little dearer than his horse.

ALFRED, LORD TENNYSON

Don't get us wrong. Darwin knew plenty, and he deserves his place in the pantheon of great thinkers. If you're a Darwin-basher looking for support, you'll find little here. Charles Darwin was a genius and a gentleman for whom we have endless respect. But as is often the case with gentleman geniuses, he was a bit clueless when it came to women.

In questions of human sexual behavior, Darwin had little to go on other than conjecture. His own sexual experience appears to have been limited to his vehemently proper wife, Emma Wedgwood, who was also his first cousin. During his circumnavigation of the globe on the *Beagle*, the young naturalist appears never to have gone ashore in search of the sexual and sensual pleasures pursued by many seafaring men of that era. Darwin was apparently far too inhibited for the decidedly hands-on data collection Herman Melville hinted at in his best-selling novels *Typee* and *Omoo*, or to sample the dusky South Pacific pleasures that had inspired the sexually frustrated crew of *The Bounty* to mutiny.

Darwin was far too buttoned-up for such carnal pursuits. His by-the-book approach to such matters is evident in his careful consideration of marriage in the abstract, before he even had any particular woman in mind. He sketched out the pros and cons in his notebook: *Marry* and *Not Marry*. On the *Marry* side he listed, "Children—(if it Please God)—Constant companion, (& friend in old age) who will feel interested in one, —object to be loved and played with. —Better than a dog anyhow . . . female chit-chat . . . *but terrible loss of time*."

On the other side of the page, Darwin listed concerns such as "Freedom to go where one liked—choice of Society & *little of it*. . . . Not forced to visit relatives & to bend in every trifle . . . fatness & idleness—Anxiety & responsibility. . . . Perhaps my wife wont [sic] like London; then the sentence is banishment & degradation into indolent, idle fool."[12]

Though Darwin proved to be a very loving husband and father, these pros and cons of marriage suggest he very seriously considered opting for the companionship of a dog instead.

The Flintstonization of Prehistory

"Judging from the social habits of man as he now exists" is anything but a reliable method for understanding prehistory (though admittedly,

Darwin had little else to go on). The search for clues to the distant past among the overwhelming detail of the immediate present tends to generate narratives closer to self-justifying myth than to science.

The word *myth* has been debased and cheapened in modern usage; it's often used to refer to something false, a lie. But this use misses the deepest function of *myth*, which is to lend narrative order to apparently disconnected bits of information, the way constellations group impossibly distant stars into tight, easily recognizable patterns that are simultaneously imaginary *and* real. Psychologists David Feinstein and Stanley Krippner explain, "Mythology is the loom on which [we] weave the raw materials of daily experience into a coherent story." This weaving becomes tricky indeed when we mythologize about the daily experience of ancestors separated from us by twenty or thirty thousand years or more. All too often, we inadvertently weave our own experiences into the fabric of prehistory. We call this widespread tendency to project contemporary cultural proclivities into the distant past "Flintstonization."[13]

Just as the Flintstones were "the modern stone-age family," contemporary scientific speculation concerning prehistoric human life is often distorted by assumptions that *seem* to make perfect sense. But these assumptions can lead us far from the path to truth.

Flintstonization has two parents: a lack of solid data and the psychological need to explain, justify, and celebrate one's own life and times. But for our purposes, Flintstonization has at least three intellectual grandfathers: Hobbes, Rousseau, and Malthus.

Thomas Hobbes (1588–1679), a lonely, frightened war refugee in Paris, was Flintstoned when he looked into the mists of prehistory and conjured miserable human lives that were "solitary, poor, nasty, brutish and short." He conjured a prehistory very much like the world he saw around him in seventeenth-century Europe, yet gratifyingly worse in every respect. Propelled by a very different psychological agenda, Jean-Jacques Rousseau (1712–1778) looked at the suffering and filth of European societies and thought he saw the corruption of a pristine human nature. Travelers' tales of simple savages in the Americas fueled his romantic fantasies. The intellectual pendulum swung back toward the Hobbesian view a few decades later when Thomas Malthus (1766–1834) claimed to mathematically demonstrate that extreme poverty and its attendant desperation typify the eternal human condition. Destitution, he argued, is intrinsic to the calculus of mammalian reproduction. As long as population increases geometrically, doubling each generation (2, 4, 8, 16, 32, etc.), and farmers can increase the food supply only by adding acreage arithmetically (1, 2, 3, 4, etc.), there will never—*can never*—be enough for everyone. Thus, Malthus concluded that poverty is as inescapable as the wind and the rain. Nobody's fault. Just the way it is. This conclusion was very popular with the wealthy and powerful, who were understandably eager to make sense of their good fortune and justify the suffering of the poor as an unavoidable fact of life.

Darwin's *eureka* moment was a gift from two terrible Thomases and one friendly Fred: Hobbes, Malthus, and Flintstone, respectively. By articulating a detailed (albeit erroneous) description of human nature and the sorts of lives humans lived in prehistory, Hobbes and Malthus provided the intellectual context for Darwin's theory of natural selection. Unfortunately, their thoroughly Flintstoned assumptions are fully integrated into Darwin's thinking and persist to the present day.

The sober tones of serious science often mask the mythical nature of what we're told about prehistory. And far too often, the myth is dysfunctional, inaccurate, and self-justifying.

Our central ambition for this book is to distinguish some of the stars from the constellations. We believe that the generally accepted myth of the origins and nature of human sexuality is not merely factually flawed, but destructive, sustaining a false sense of what it means to be a human being. This false narrative distorts our sense of our capacities and needs. It amounts to false advertising for a garment that fits almost no one. But we're all supposed to buy and wear it anyway.

Like all myths, this one seeks to define who and what we are and thus what we can expect and demand from one another. For centuries, religious authorities disseminated this defining narrative, warning of chatty serpents, deceitful women, forbidden knowledge, and eternal agony. But more recently, it's been marketed to secular society as hard science.

Examples abound. Writing in the prestigious journal *Science*, anthropologist Owen Lovejoy suggested, "The nuclear family and human sexual behavior may have their ultimate origin long before the dawn of the Pleistocene [1.8 million years ago]."[14] Well-known anthropologist Helen Fisher concurs, writing, "Is monogamy natural?" She gives a one-word answer: "Yes." She then continues, "Among human beings . . . monogamy is the rule."[15]

Many different elements of human prehistory seem to nest neatly into each other in the standard narrative of human sexual evolution. But remember, that Indian *seemed* to answer Cortés's question, and it *seemed* indisputable to Pope Urban VIII and just about everyone else that the Earth remained solidly at the center of the solar system. With a focus on the presumed nutritional benefits of pair-bonding, zoologist and science writer Matt Ridley demonstrates the seduction in this apparent unity: "Big brains needed meat . . . [and] food sharing allowed a meaty diet (because it freed men to risk failure in pursuit of game) . . . [and] food sharing demanded big brains (without detailed calculating memories, you could easily be cheated by a free-loader)." So far, so good. But now Ridley inserts the sexual steps in his dance: "The sexual division of labor promoted monogamy (a pair bond now being an economic unit); monogamy led to neotenous sexual selection (by putting a premium on youthfulness in mates)." It's a waltz, with one assumption spinning into the next, circling round and round in "a spiral of comforting justification, proving how we came to be as we are."[16]

Note how each element anticipates the next, all coming together in a tidy constellation that seems to explain human sexual evolution.

The distant stars fixed in the standard constellation include:

- what motivated prehuman males to "invest" in a particular female and her children;
- male sexual jealousy and the double standard concerning male versus female sexual autonomy;
- the oft-repeated "fact" that the timing of women's ovulation is "hidden";
- the inexplicably compelling breasts of the human female;
- her notorious deceptiveness and treachery, source of many country and blues classics;
- and of course, the human male's renowned eagerness to screw anything with legs—an equally rich source of musical material.

This is what we're up against. It's a song that is powerful, concise, self-reinforcing, and playing on the radio all day and all night . . . but still wrong, baby, oh so wrong.

The standard narrative is about as scientifically valid as the story of Adam and Eve. In many ways, in fact, it is a scientific retelling of the Fall into original sin as depicted in *Genesis*—complete with sexual deceit, prohibited knowledge, and guilt. It hides the truth of human sexuality behind a fig leaf of anachronistic Victorian discretion repackaged as science. But actual—as opposed to mythical—science has a way of peeking out from behind the fig leaf.

Charles Darwin proposed two basic mechanisms through which evolutionary change occurs. The first, and better known, is *natural selection*. Economic philosopher Herbert Spencer later coined the phrase "survival of the fittest" to describe this mechanism, though most biologists still prefer "natural selection." It's important to understand that evolution is *not* a process of improvement. Natural selection simply asserts that species change as they adapt to ever-changing environments. One of the chronic mistakes made by would-be social Darwinists is to assume that evolution is a process by which human beings or societies become *better*.[17] It is not.

Those organisms best able to survive in a challenging, shifting environment live to reproduce. As survivors, their genetic code likely contains information advantageous to their offspring in *that particular environment*. But the environment can change at any moment, thus neutralizing the advantage.

Charles Darwin was far from the first to propose that some sort of evolution was taking place in the natural world. Darwin's grandfather, Erasmus Darwin, had noted the process of differentiation evident in both plants and animals. The big question was *how* it happened: What was the mechanism by which species differentiated from each other? Darwin was particularly struck by the subtle differences in the finches he'd seen on various islands in the Galápagos. This insight suggested that environment was crucial to the process, but until later, he had no way to explain *how* the environment shapes organisms over generations.

What Is Evolutionary Psychology and Why Should You Care?

Evolutionary theory has been applied to the body pretty much since Darwin published *On the Origin of Species*. He'd been sitting on his theory for decades, fearing the controversy sure to follow its publication. If you want to know why human beings have ears on the sides of their heads and eyes up front, evolutionary theory can tell you, just as it can tell you why birds have their eyes on the sides of their heads and no visible ears at all. Evolutionary theory, in other words, offers explanations of how *bodies* came to be as they are.

In 1975, E. O. Wilson made a radical proposal. In a short, explosive book called *Sociobiology*, Wilson argued that evolutionary theory could be, indeed *must* be, applied to behavior—not just bodies. Later, to avoid rapidly accumulating negative connotations—some associated with eugenics (founded by Darwin's cousin, Francis Galton)—the approach was renamed "evolutionary psychology" (EP). Wilson proposed to bring evolutionary theory to bear on a few "central questions . . . of unspeakable importance: How does the mind work, and beyond that why does it work in such a way and not another, and from these two considerations together, what is man's ultimate nature?" He argued that evolutionary theory is "the essential first hypothesis for any serious consideration of the human condition," and that "without it the humanities and social sciences are the limited descriptors of surface phenomena, like astronomy without physics, biology without chemistry, and mathematics without algebra."[18]

Beginning with *Sociobiology*, and *On Human Nature*, a follow-up volume Wilson published three years later, evolutionary theorists began to shift their focus from eyes, ears, feathers, and fur to less tangible, far more contentious issues such as love, jealousy, mate choice, war, murder, rape, and altruism. Juicy subject matter lifted from epics and soap operas became fodder for study and debate in respectable American universities. Evolutionary psychology was born.

It was a difficult birth. Many resented the implication that our thoughts and feelings are as hard-wired in our genetic code as the shape of our heads or the length of our fingers—and thus presumably as inescapable and unchangeable. Research in EP quickly became focused on differences between men and women, shaped by their supposedly conflicting reproductive agendas. Critics heard overtones of racial determinism and the smug sexism that had justified centuries of conquest, slavery, and discrimination.

Although Wilson never argued that genetic inheritance *alone* creates psychological phenomena, merely that evolved tendencies *influence* cognition and behavior, his moderate insights were quickly obscured by the immoderate disputes they sparked. Many social scientists at the time believed humans to be nearly completely cultural creatures, blank slates to be marked by society.[19] But Wilson's perspective was highly attractive to other academics eager to introduce a more rigorous scientific methodology into fields they considered overly subjective and distorted by liberal political views and wishful thinking. Decades later, the two sides of the debate remain largely entrenched in their extreme positions: human behavior as genetically determined versus human behavior as socially determined. As you might expect, the truth—and the most valuable science being done in the field—lies somewhere in between these two extremes.

Today, self-proclaimed EP "realists" argue that it's ancient human nature that leads us to wage war on our neighbors, deceive our spouses, and abuse our stepchildren. They argue that rape is an unfortunate, but largely successful reproductive strategy and that marriage amounts to a no-win struggle of mutually assured disappointment. Romantic love is reduced to a chemical reaction luring us into reproductive entanglements parental love keeps us from escaping. Theirs is an all-encompassing narrative claiming to explain it all by reducing every human interaction to the reptilian pursuit of self-interest.[20]

Of course, there are many scientists working in evolutionary psychology, primatology, evolutionary biology, and other fields who don't sign on to the narrative we're critiquing in these pages, or whose paradigms overlap at some points but differ at others. We hope they'll forgive us if it sometimes seems we oversimplify in order to more clearly illustrate the broad outlines of the various paradigms without getting lost in the weeds of subtle differences. (Readers seeking more detailed information are encouraged to consult the endnotes.)

Evolutionary psychology's standard narrative contains several clanging contradictions, but one of the most discordant involves female libido. Females, we're told again and again, are the choosy, reserved sex. Men spend their energies trying to impress women—flaunting expensive watches, packaging themselves in shiny new sports cars, clawing their way to positions of fame, status, and power—all to convince coy females to part with their closely guarded sexual favors. For women, the narrative holds that sex is about the security—emotional and material—of the relationship, not the physical pleasure. Darwin agreed with this view. The "coy" female who "requires to be courted" is deeply embedded in his theory of sexual selection.

If women were as libidinous as men, we're told, society itself would collapse. Lord Acton was only repeating what everyone knew in 1875 when he declared, "The majority of women, happily for them and for society, are not very much troubled with sexual feeling of any kind."

And yet, despite repeated assurances that women aren't particularly sexual creatures, in cultures around the world men have gone to extraordinary lengths to control female libido: female genital mutilation, head-to-toe chadors, medieval witch burnings, chastity belts, suffocating corsets, muttered insults about "insatiable" whores, pathologizing, paternalistic medical diagnoses of nymphomania or hysteria, the debilitating scorn heaped on any female who chooses to be generous with her sexuality . . . all parts of a worldwide campaign to keep the supposedly low-key female libido under control. Why the electrified high-security razor-wire fence to contain a kitty-cat?

Tiresias, a prominent figure in Greek mythology, had a unique perspective on male and female sexual pleasure.

While still a young man, Tiresias came upon two snakes entwined in copulation. With his walking stick, he separated the amorous serpents and was suddenly transformed into a woman.

Seven years later, the female Tiresias was walking through the forest when she again interrupted two snakes in a private moment. Placing her staff between them, she completed the cycle and was transformed back into a man.

This unique breadth of experience led the first couple of the Greek pantheon, Zeus and Hera, to call upon Tiresias to resolve their long-running marital dispute: who enjoys sex more, men or women? Zeus was sure that women did, but Hera would hear none of it. Tiresias replied that not only did females enjoy sex more than males, they enjoyed it nine times more!

His response incensed Hera so much that she struck Tiresias blind. Feeling responsible for having dragged poor Tiresias into this mess, Zeus tried to make amends by giving him the gift of prophesy. It was from this state of blinded vision that Tiresias saw the terrible destiny of Oedipus, who unknowingly killed his father and married his mother.

Peter of Spain, author of one of the most widely read medical books of the thirteenth century, the *Thesaurus Pauperum*, was more diplomatic when confronted with the same question. His answer (published in *Quaestiones super Viaticum*) was that although it was true women experienced greater *quantity* of pleasure, men's sexual pleasure was of higher *quality*. Peter's book included ingredients for thirty-four aphrodisiacs, fifty-six prescriptions to enhance male libido, and advice for women wanting to avoid pregnancy. Perhaps it was his diplomacy, the birth-control advice, or his open-mindedness that led to one of history's strange and tragic turns. In 1276, Peter of Spain was elected Pope John XXI, but he died just nine months later when the ceiling of his library suspiciously collapsed on him as he slept.

Why does any of this history matter? Why is it important that we correct widely held misconceptions about human sexual evolution?

Well, ask yourself what might change if everyone knew that women do (or, at least, can, in the right circumstances) enjoy sex as much as men, not to mention nine times more, as Tiresias claimed? What if Darwin was wrong about the sexuality of the human female—led astray by his Victorian bias? What if Victoria's biggest secret was that men and women are *both* victims of false propaganda about our true sexual natures and *the war between the sexes*—still waged today—is a false-flag operation, a diversion from our common enemy?

We're being misled and misinformed by an unfounded yet constantly repeated mantra about the *naturalness* of wedded bliss, female sexual reticence, and happily-ever-after sexual monogamy—a narrative pitting man against woman in a tragic tango of unrealistic expectations, snowballing frustration, and crushing disappointment. Living under this *tyranny of two*, as author and media critic Laura Kipnis puts it, we carry the weight of "modern love's central anxiety," namely, "the expectation that romance and sexual attraction can last a lifetime of coupled togetherness despite much hard evidence to the contrary."[21]

We build our most sacred relationships on the battleground where evolved appetites clash with the romantic mythology of monogamous marriage. As Andrew J. Cherlin recounts in *The Marriage-Go-Round*, this unresolved conflict between what we are and what many wish we were results in "a great turbulence in American family life, a family flux, a coming and going of partners on a scale seen nowhere else." Cherlin's research shows that "[t]here are more partners in the personal lives of Americans than in the lives of people of any other Western country."[22]

But we rarely dare to confront the contradiction at the heart of our mistaken ideal of marriage head-on. And if we do? During a routine discussion of yet another long-married politician caught with his pants down, comedian/social critic Bill Maher asked the guests on his TV show to consider the unspoken reality underlying many of these situations: "When a man's been married twenty years," Maher said, "he doesn't want to have sex, or his wife doesn't want to have sex with him. Whatever it is. *What is the right answer?* I mean, I know he's bad for cheating, but *what's the right answer?* Is it—to just suck it up and live the rest of your life passionless, and imagine somebody else when you're having sex with your wife the three days a year that you have sex?" After an extended, awkward silence, one of Maher's panelists eventually suggested, "The right answer is to get out of the relationship. . . . Move on. I mean, you're an adult." Another agreed, noting, "Divorce is legal in this country." The third, normally outspoken journalist P. J. O'Rourke, just looked down at his shoes and said nothing.

"Move on?" Really? Is abandonment of one's family the "adult" option for dealing with the inherent conflict between socially sanctioned romantic ideals and the inconvenient truths of sexual passion?[23]

Darwin's sense of the *coy female* wasn't based only on his Victorian assumptions. In addition to natural selection, he proposed a second mechanism for evolutionary change: *sexual selection.* The central premise of sexual selection is that in most mammals, the female has a much higher investment in offspring than does the male. She's stuck with gestation, lactation, and extended nurturing of the young. Because of this inequality in unavoidable sacrifice, Darwin reasoned, she is the more hesitant participant, needing to be convinced it's a good idea—while the male, with his slam-bam-thank-you-ma'am approach to reproduction, is eager to do the convincing. Evolutionary psychology is founded on the belief that male and female approaches to mating have intrinsically conflicted agendas.

The selection of the winning bachelor typically involves male competition: rams slamming their heads together, peacocks dragging around colorful, predator-attracting tails, men bearing expensive gifts and vowing eternal love over candlelight. Darwin saw sexual selection as a struggle between males for sexual access to passive, fertile females who would submit to the victor. Given the competitive context his theories assume, he believed "promiscuous intercourse in a state of nature [to be] extremely improbable." But at least one of Darwin's contemporaries disagreed.

Lewis Henry Morgan

To white people, he was known as Lewis Henry Morgan (1818–1881), a railroad lawyer with a fascination for scholarship and the ways in which societies organize themselves.[24] The Seneca tribe of the Iroquois

Nation adopted Morgan as an adult, giving him the name *Tayadaowuh-kuh*, which means "bridging the gap." At his home near Rochester, New York, Morgan spent his evenings studying and writing, trying to bring scientific rigor to understanding the intimate lives of people made distant by time or space. The only American scholar to have been cited by each of the other three intellectual giants of his century, Darwin, Freud, and Marx, many consider Morgan the most influential social scientist of his era and the father of American anthropology. Ironically, it may be Marx and Engels's admiration that explains why Morgan's work isn't better known today. Though he was no Marxist, Morgan doubted important Darwinian assumptions concerning the centrality of sexual competition in the human past. This stance was enough to offend some of Darwin's defenders—though not Darwin himself, who respected and admired Morgan. In fact, Morgan and his wife spent an evening with the Darwins during a trip to England. Years later, two of Darwin's sons stayed with the Morgans at their home in upstate New York.

Morgan was especially interested in the evolution of family structure and overall social organization. Contradicting Darwinian theory, he hypothesized a far more promiscuous sexuality as having been typical of prehistoric times. "The husbands lived in polygyny [i.e., more than one wife], and the wives in polyandry [i.e., more than one husband], which are seen to be as ancient as human society. Such a family was neither unnatural nor remarkable," he wrote. "It would be difficult to show any other possible beginning of the family in the primitive period." A few pages later Morgan concludes that "there seems to be no escape" from the conclusion that a "state of promiscuous intercourse" was typical of prehistoric times, "although questioned by so eminent a writer as Mr. Darwin."[25]

Morgan's argument that prehistoric societies practiced group marriage (also known as *the primal horde* or *omnigamy*—the latter term apparently coined by French author Charles Fourier) so influenced Darwin's thinking that he admitted, "It seems certain that the habit of marriage has been gradually developed, and that almost promiscuous intercourse was once extremely common throughout the world." With his characteristic courteous humility, Darwin agreed that there were "present day tribes" where "all the men and women in the tribe are husbands and wives to each other." In deference to Morgan's scholarship, Darwin continued, "Those who have most closely studied the subject, and whose judgment is worth much more than mine, believe that communal marriage was the original and universal form throughout the world. . . . The indirect evidence in favour of this belief is extremely strong. . . ."[26]

Indeed it is. And the evidence—both direct and indirect—has grown much stronger than Darwin, or even Morgan, could have imagined.

But first, a word about a word. *Promiscuous* means different things to different people, so let's define our terms. The Latin root is *miscere*, "to mix," and that's how we mean it. We don't imply any *randomness* in mating, as choices and preferences still exert their influence. We looked for another term to use in this book, one without the derogatory sneer, but the synonyms are even worse: *sluttish, wanton, whorish, fallen.*

Please remember that when we describe the sexual practices in various societies around the world, we're describing behavior that is *normal* to the people in question. In the common usage, *promiscuity* suggests immoral or amoral behavior, uncaring and unfeeling. But most of the people we'll be describing are acting well within the bounds of what their society considers acceptable behavior. They're not rebels, transgressors, or utopian idealists. Given that groups of foragers (either those still existing today or in prehistoric times) rarely number much over 100 to 150 people, each is likely to know every one of his or her partners deeply and intimately—probably to a much greater degree than a modern man or woman knows his or her casual lovers.

Morgan made this point in *Ancient Society*, writing, "This picture of savage life need not revolt the mind, because to them it was a form of the marriage relation, and therefore devoid of impropriety."[27]

Biologist Alan F. Dixson, author of the most comprehensive survey of primate sexuality (called, unsurprisingly, *Primate Sexuality*), makes a similar point concerning what he prefers to call "multimale-multifemale mating systems" typical of our closest primate relations: chimps and bonobos. He writes, "Mating is rarely indiscriminate in multimale-multifemale primate groups. A variety of factors, including kinship ties, social rank, sexual attractiveness and individual sexual preferences might influence mate choice in both sexes. It is, therefore, incorrect to label such mating systems as *promiscuous*."[28]

So, if *promiscuity* suggests a number of ongoing, nonexclusive sexual relationships, then yes, our ancestors were far more promiscuous than all but the randiest among us. On the other hand, if we understand *promiscuity* to refer to a lack of discrimination in choosing partners or having sex with random strangers, then our ancestors were likely far *less* promiscuous than many modern humans. For this book, *promiscuity* refers only to having a number of ongoing sexual relationships at the same time. Given the contours of prehistoric life in small bands, it's unlikely that many of these partners would have been strangers.

CHAPTER THREE

A Closer Look at the Standard Narrative of Human Sexual Evolution

We have good news and bad news. The good news is that the dismal vision of human sexuality reflected in the standard narrative is mistaken. Men have *not* evolved to be deceitful cads, nor have millions of years shaped women into lying, two-timing gold-diggers. But the bad news is that the amoral agencies of evolution have created in us a species with a secret it just can't keep. *Homo sapiens* evolved to be shamelessly, undeniably, inescapably sexual. Lusty libertines. Rakes, rogues, and roués. Tomcats and sex kittens. Horndogs. Bitches in heat.[1]

True, some of us manage to rise above this aspect of our nature (or to sink below it). But these preconscious impulses remain our biological baseline, our reference point, the *zero* in our own personal number system. Our evolved tendencies are considered "normal" by the body each of us occupies. Willpower fortified with plenty of guilt, fear, shame, and mutilation of body and soul may provide some control over these urges and impulses. Sometimes. Occasionally. Once in a blue moon. But even when controlled, they refuse to be ignored. As German philosopher Arthur Schopenhauer pointed out, *Mensch kann tun was er will; er kann aber nicht wollen was er will.* (One can choose what to do, but not what to want.)

Acknowledged or not, these evolved yearnings persist and clamor for our attention.

And there are costs involved in denying one's evolved sexual nature, costs paid by individuals, couples, families, and societies every day and every night. They are paid in what E. O. Wilson called "the less tangible currency of human happiness that must be spent to circumvent our natural predispositions."[2] Whether or not our society's investment in sexual repression is a net gain or loss is a question for another time. For now, we'll just suggest that trying to rise above nature is always a risky, exhausting endeavor, often resulting in spectacular collapse.

Any attempt to understand who we are, how we got to be this way, and what to do about it must begin by facing up to our evolved human sexual predispositions. Why do so many forces resist our sustained fulfillment? Why is conventional marriage so much damned work? How has the incessant, grinding campaign of socio-scientific insistence upon the *naturalness* of sexual monogamy combined with a couple thousand years of fire and brimstone failed to rid even the priests, preachers, politicians, and professors of their prohibited desires? To see ourselves as we are, we must begin by acknowledging that of all Earth's creatures, none is as urgently, creatively, and constantly sexual as *Homo sapiens*.

We don't claim that men and women experience their eroticism in precisely the same ways, but as Tiresias noted, both women and men find considerable pleasure there. True, it may take most women a bit longer to get the sexual motor running than it does men, but once warmed up, most women are fully capable of leaving any man far behind. No doubt, males tend to be more concerned with a woman's looks, while most women find a man's character more compelling than his appearance (within limits, of course). And it's true that women's biology gives them a lot more to consider before a roll in the hay.

Comedian Jerry Seinfeld sums it up in terms of fire and firemen: "The basic conflict between men and women, sexually, is that men are like firemen. To men, sex is an emergency, and no matter what we're doing we can be ready in two minutes. Women, on the other hand, are like fire. They're very exciting, but the conditions have to be exactly right for it to occur."

Perhaps for many women libido is like the hunger of a gourmand. Unlike many men, such women don't yearn to eat just to stop the hunger. They're looking for particular satisfactions presented in certain ways. Where most men can and do hunger for sex in the abstract, women report wanting narrative, character, a *reason* for sex.* In other words, we agree with many of the *observations* central to evolutionary psychology—it's the contorted, internally conflicted *explanations* for these observations that we find problematic.

Still, there *are* simple, logical, consistent explanations for most of these standard observations concerning human sexuality—explanations that offer an alternative narrative of human sexual evolution that is both parsimonious and elegant; a revised model that requires none of the convoluted mixed strategies and Flintstonizing intrinsic to the currently accepted story.

The standard narrative paints a dark image of our species over a much brighter—albeit somewhat scandalous—truth. Before presenting our model in detail, let's take a closer look at the standard narrative, focusing on the four major areas of research that incorporate the most widely accepted assumptions:

- The relatively weak female libido
- Male parental investment (MPI)
- Sexual jealousy and paternity certainty
- Extended receptivity and concealed (or cryptic) ovulation

How Darwin Insults Your Mother (The Dismal Science

* But who would argue the gourmand takes *less* pleasure in her food than the glutton?

of Sexual Economics)

What does the winning male suitor supposedly get for all his preening and showing off? Sex. Well, not just sex, but *exclusive* access to a particular woman. The standard model posits that sexual exclusivity is crucial because in evolutionary times this was a man's only way of ensuring his paternity. According to evolutionary psychology, this is the grudging agreement at the heart of the human family. Men offer goods and services (in prehistoric environments, primarily meat, shelter, protection, and status) in exchange for exclusive, relatively consistent sexual access. Helen Fisher called it *The Sex Contract.*

Economics, often referred to as *the dismal science*, is never more dismal than when applied to human sexuality. The sex contract is often explained in terms of economic game theory in which she or he who has the most offspring surviving to reproduce wins—because her or his "return on investment" is highest. So, if a woman becomes pregnant by a guy who has no intention of helping her through pregnancy or guiding the child through the high-risk early years, she likely is squandering the time, energy, and risks of pregnancy. According to this theory, without the help of the father, chances are much better that the child will die before reaching sexual maturity—not to mention the increased health risks to the pregnant or nursing mother. Prominent evolutionary psychologist Steven Pinker calls this way of looking at human reproduction *the genetic economics of sex*: "The minimum investments of a man and a woman are . . . unequal," explains Pinker, "because a child can be born to a single mother whose husband has fled but not to a single father whose wife has fled. But the investment of the man is greater than zero, which means that women are also predicted to compete in the marriage market, though they should compete over the males most likely to invest . . ."[3]

Conversely, if a guy invests all his time, energy, and resources in a woman who's doing the nasty behind his back, he's at risk of raising another man's kids—a total loss if his sole purpose in life is getting his own genes into the future. And make no mistake: according to the cold logic of standard evolutionary theory, leaving a genetic legacy *is* our sole purpose in life. This is why evolutionary psychologists Margo Wilson and Martin Daly argue that men take a decidedly *proprietary* view of women's sexuality: "Men lay claim to particular women as songbirds lay claim to territories, as lions lay claim to a kill, or as people of both sexes lay claim to valuables," they write. "Having located an individually recognizable and potentially defensible resource packet, the proprietary creature proceeds to advertise and exercise the intention of defending it from rivals."[4]

"Baby, I love you like a lion loves his kill." Surely, a less romantic description of marriage has never been written.

As attentive readers may have noted, the standard narrative of heterosexual interaction boils down to prostitution: a woman exchanges her sexual services for access to resources. Maybe mythic resonance explains part of the huge box-office appeal of a film like *Pretty Woman*, where Richard Gere's character trades access to his wealth in exchange for what Julia Roberts's character has to offer (she plays a hooker with a heart of gold, if you missed it). Please note that what she's got to offer is limited to the aforementioned heart of gold, a smile as big as Texas, a pair of long, lovely legs, and the solemn promise that they'll open only for him from now on. The genius of *Pretty Woman* lies in making explicit what's been implicit in hundreds of films and books. According to this theory, women have evolved to unthinkingly and unashamedly exchange erotic pleasure for access to a man's wealth, protection, status, and other treasures likely to benefit her and her children.

Darwin says your mother's a whore. Simple as that.

Lest you think we're being flip, we assure you that the bartering of female fertility and fidelity in exchange for goods and services is one of the foundational premises of evolutionary psychology. *The Adapted Mind*, a book many consider to be the bible of the field, spells out the sex contract very clearly:

> A man's sexual attractiveness to women will be a function of traits that were correlated with high mate value in the natural environ-

ment. . . . The crucial question is, What traits would have been correlated with high mate value? Three possible answers are as follows:

- The willingness and ability of a man to provide for a woman and her children. . . .
- The willingness and ability of a man to protect a woman and her children. . . .
- The willingness and ability of a man to engage in direct parenting activities.[5]

Now let's review some of the most prominent research founded upon these assumptions about men, women, family structure, and prehistoric life.

The Famously Flaccid Female Libido

The female . . . with the rarest exception, is less eager than the male. . . .

CHARLES DARWIN

Women have little interest in sex, right? Despite Tiresias's observations, until very recently, that's been the near-universal consensus in Western popular culture, medicine, and evolutionary psychology. In recent years, popular culture has begun to question women's relative lack of interest, but as far as the standard model is concerned, not much has changed since Dr. William Acton published his famous thoughts on the matter in 1875, assuring his readers, "The best mothers, wives, and managers of households know little or nothing of sexual indulgences. . . . As a general rule, a modest woman seldom desires any sexual gratification for herself. She submits to her husband, but only to please him."[6]

More recently, in his now classic work *The Evolution of Human Sexuality*, psychologist Donald Symons confidently proclaimed that "among all peoples sexual intercourse is understood to be a service or favor that females render to males."[7] In a foundational paper published in 1948, geneticist A. J. Bateman wasn't hesitant to extrapolate his findings concerning fruit fly behavior to humans, commenting that natural selection encourages "an undiscriminating eagerness in the males and a discriminating passivity in the females."[8]

The sheer volume of evidence amassed to convince us that women are not particularly sexual beings is quite impressive. Hundreds, if not thousands, of studies have claimed to confirm the flaccidity of the female libido. One of the most cited studies in all of evolutionary psychology, published by 1989, is typical of the genre.[9] An attractive undergraduate student volunteer walked up to an unsuspecting student of the opposite sex (who was alone) on the campus of Florida State University and said, "Hi, I've been noticing you around town lately, and I find you very attractive. Would you go to bed with me tonight?" About 75 percent of the young men said yes. Many of those who didn't asked for a "rain check." But *not one* of the women approached by these attractive strangers accepted the offer. Case closed.

Seriously, this study *really is* one of the best known in all of EP. Researchers reference it to establish that women aren't interested in casual sex, which is important if your theory posits that women instinctively barter sex to get things from men. After all, if they're giving it away for free, the bottom falls out of the market, and other women are going to have a harder time exchanging sex for anything of value.

Male Parental Investment (MPI)

As mentioned above, underlying each of these theories, as well as evolutionary theory in general, is the notion that life can be conceptualized in terms of economics and game theory. The objective of the game is to send your genetic code into the future by producing the maximum possible number of offspring who survive and reproduce. Whether or not this dispersal leads to happiness is irrelevant. In his best-selling survey of EP, *The Moral Animal*, Robert Wright puts it succinctly, saying: "We are built to be effective animals, not happy ones. (Of course, we're designed to pursue happiness; and the attainment of Darwinian goals—sex, status, and so on—often brings happiness, at least for a while.) Still, the frequent absence of happiness is what keeps us pursuing it, and thus makes us productive."[10]

This is a curious notion of productivity—at once overtly political and yet presented innocently enough, as if there were only one possible meaning of "productivity." This perspective on life incorporates the Protestant work ethic (that "productivity" is what makes an animal "effective") and echoes the *Old Testament* notion that life must be endured, not enjoyed. These assumptions are embedded throughout the literature of evolutionary psychology. Ethologist/primatologist Frans de Waal, one of the more open-minded philosophers of human nature, calls this *Calvinist sociobiology.*

The female interest in quality over quantity is thought to be important in two respects. First, she would clearly be interested in conceiving a child with a healthy man, so as to maximize the odds that her child would survive and prosper. "Women's reproductive resources are precious and finite, and ancestral women did not squander them on just any random man," writes evolutionary psychologist David Buss. "Obviously, women don't consciously think that sperm are cheap and eggs are expensive," Buss continues, "but women in the past who failed to exercise acumen before consenting to sex were left in the evolutionary dust; our ancestral mothers used emotional wisdom to screen out losers."[11] Buss doesn't explain why there are still so many "losers" in the gene pool today if their ancestors were subject to such careful screening for thousands of generations.

While a substantial amount of female parental investment is biologically unavoidable in our species, evolutionary theorists believe that *Homo sapiens* is uniquely high in *male parental investment* (MPI) among primates. They argue that our high level of MPI forms the basis for the supposed universality of marriage. As Wright puts it, "In every human culture in the anthropological record, marriage . . . is the norm, and the family is the atom of social organization. Fathers everywhere feel love for their children. . . . This love leads fathers to help feed and defend their children, and teach them useful things."[12]

Biologist Tim Birkhead agrees, writing, "The issue of paternity is at the core of much of men's behaviour—and for good evolutionary reasons. In our primeval past, men who invested in children which were not their own would, on average, have left fewer descendents than those who reared only their own genetic offspring. As a consequence men were, and continue to be, preoccupied with paternity. . . ."[13]

For now, we'll briefly note a few of the questionable assumptions underlying this argument:

- *Every* culture is organized around marriage and the nuclear family.
- Human fathers that provided for *only* their own children would have left far more descendants than those less selective in their material generosity.
  - *Note how this presumes a discrete genetic basis for something as amorphous as "preoccupation with paternity."*
- In the ancestral environment, a man could know which children were biologically his, which presumes that:
  - *he understands that one sex act can lead to a child, and*
  - *he has 100 percent certainty of his partner's fidelity.*
- A hunter could refuse to share his catch with other hungry people living in the close-knit band of foragers (including nieces, nephews, and children of lifelong friends) without being shamed, shunned, and banished from the community.

So, according to the standard narrative, as male parental investment translates into advantages for that man's children (more food, protection, and education—other kids be damned), women would have evolved to choose mates with access to more of these resources and whose behavior indicated that they would share these resources only with her and her children (indications of selective generosity, fidelity, and sincerity).

But, according to this narrative, these two female objectives (good genes and access to a male's resources) create conflictive situations for men and women—both within their relationship and with their same-sex competitors. Wright summarizes this understanding of the situation: "High male parental investment makes sexual selection work in two directions at once. Not only have males evolved to compete for scarce female eggs; females have evolved to compete for scarce male investment."[14]

"Mixed Strategies" in the War Between the Sexes

It's no accident that the man who famously observed that power is the greatest aphrodisiac was not, by a long shot, good-looking.[15] Often (in what we might call the *Kissinger effect*), the men with the greatest access to resources and status lack the genetic wealth signified by physical attractiveness. What's a girl to do?

Conventional theory suggests she'll marry a nice, rich, predictable, sincere guy likely to pay the mortgage, change the diapers, and take out the trash—but then cheat on him with wild, sexy, dangerous dudes, especially around the time she's ovulating, so she's more likely to have lover-boy's baby. Known as the *mixed strategy* in the scientific literature, both males and females are said to employ their own version of the dark strategy in keeping with their opposed objectives in mating (females maximizing quality of mates and males maximizing quantity of mating opportunities). It's a jungle out there.

The best-known studies purporting to demonstrate the nature of these two differing strategies are those done by David Buss and his colleagues. Their hypothesis holds that if males and females have conflicting agendas concerning mating behavior, the differences should appear in the ways males and females experience sexual jealousy. These researchers found that women were consistently more upset by thoughts of their mates' *emotional* infidelity, while men showed more anxiety concerning their mates' *sexual* infidelity, as the hypothesis predicts.

These results are often cited as confirmation of the male parental investment–based model. They appear to reflect the differing interests the model predicts. A woman, according to the theory, would be more upset about her partner's emotional involvement with another woman, as that would threaten her vital interests more. According to the standard model, the worst-case scenario for a prehistoric woman in this evolutionary game would be to lose access to her man's resources and support. If he limits himself to a meaningless sexual dalliance with another woman (in modern terms, preferably a woman of a lower social class or a prostitute—whom he would be unlikely to marry), this would be far less threatening to her standard of living and that of her children. However, if he were to fall in love with another woman and leave, the woman's prospects (and those of her children) would plummet.

From the man's perspective, as noted above, the worst-case scenario would be to spend his time and resources raising another man's children (and propelling someone else's genes into the future at the expense of his own). If his partner were to have an emotional connection with another man, but no sex, this genetic catastrophe couldn't happen. But if she were to have sex with another man, even if no emotional intimacy were involved, he could find himself unknowingly losing his evolutionary "investment." Hence, the narrative predicts—and the research seems to confirm—that his jealousy should have evolved to control her *sexual* behavior (thus assuring paternity of the children), while her jealousy should be oriented toward controlling his *emotional* behavior (thus protecting her exclusive access to his resources).*

As you might guess, the *mixed strategy* referred to earlier would follow similar lines. The male's mixed strategy would be to have a long-term mate, whose sexual behavior he could control—keeping her

* We examine the nature of sexual jealousy in more detail in Chapter 9.

barefoot and pregnant if poor, foot-bound and pregnant if Chinese, or in high heels and pregnant if rich. Meanwhile he should continue having casual (low-investment) sex with as many other women as possible, to increase his chances of fathering more children. This is how standard evolutionary theory posits that men evolved to be dirty, lying bastards. According to the standard narrative, the evolved behavioral strategy for a man is to cheat on his pregnant wife while being insanely—even violently—jealous of her.

Charming.

Although the survival odds of any children resulting from his casual encounters would presumably be lower than those of the children he helps raise, this investment would still be wise for him, given the low costs he incurs (a few drinks and a room at the Shady Grove Motor Lodge—at the hourly rate). The woman's mixed strategy would be to extract a long-term commitment from the man who offers her the best access to resources, status, and protection, while still seeking the occasional fling with rugged dudes in leather jackets who offer genetic advantages her loving, but domesticated, mate lacks. It's hard to decide who comes out looking worse.

Various studies have demonstrated that women are more likely to cheat on their husbands (to have extra-pair copulations, or EPCs) when they are ovulating and less likely to use birth control than they are when not fertile. Furthermore, women are likely to wear more perfume and jewelry when ovulating than at other points in their menstrual cycle and to be attracted to more macho-looking men (those with physical markers of more vigorous genes). These conflicting agendas and the eternal struggle they appear to fuel—this "war between the sexes"—is central to the dismal vision of human sexual life featured in today's scientific and therapeutic narratives.

As Wright summarizes, "Even with high MPI [male parental investment], and in some ways because of it, a basic underlying dynamic between men and women is *mutual exploitation*. They seem, at times, *designed to make each other miserable* [emphasis added]."[16] Symons voices the same resignation in the first lines of *The Evolution of Human Sexuality*:

> A central theme of this book is that, with respect to sexuality, there is a female human nature and a male human nature, and that these natures are extraordinarily different, though the differences are to some extent masked by the compromises heterosexual relations entail and by moral injunctions. Men and women differ in their sexual natures because throughout the immensely long hunting and gathering phase of human evolutionary history the sexual desires and dispositions that were adaptive for either sex were for the other tickets to reproductive oblivion.[17]

Bleak, no? Conventional evolutionary theory assures us that all you scheming, gold-digging women reading this are evolved to trick a trusting yet boring guy into marrying you, only to then spray on a bunch of perfume and run down to the local singles club to try to get pregnant by some unshaven Neanderthal as soon as hubby falls asleep on the couch. How could you? But before male readers start feeling superior, remember that according to the same narrative, you evolved to woo and marry some innocent young beauty with empty promises of undying love, fake Rolex prominent on your wrist, get her pregnant ASAP, then start "working late" with as many secretaries as you can manage. Nothing to be proud of, mister.

Extended Sexual Receptivity and Concealed Ovulation

Unlike her closest primate cousins, the standard human female doesn't come equipped with private parts that swell up to double their normal size and turn bright red when she is about to ovulate. In fact, a foundational premise of the standard narrative is that men have no way of knowing when a woman is fertile. As we're supposed to be the smartest creatures around, it's interesting that humans are thought to be almost unique in this ignorance. The vast majority of other female mammals advertise when they are fertile, and are decidedly not interested in sex at other times. Concealed ovulation is said to be a significant human exception. Among primates, the female capacity and willingness to have sex any time, any place is characteristic only of bonobos and humans. "Extended receptivity" is just a scientific way of saying that women can be sexually active throughout their menstrual cycle, whereas most mammals have sex only when it "matters"—that is, when pregnancy can occur.

If we accept the assumption that women are not particularly interested in sex, other than as a way to manipulate men into sharing resources, why would human females have evolved this unusually abundant sexual capacity? Why not reserve sex for those few days in the cycle when pregnancy is most probable, as does practically every other mammal?

Two principal theories have been proposed to explain this phenomenon, and they couldn't be more different. What anthropologist Helen Fisher has called "the classic explanation" goes like this: both concealed ovulation and extended (or, more accurately, constant) sexual receptivity evolved among early human females as a way of developing and cementing the pair bond by holding the attention of a constantly horny male mate. This capacity supposedly worked in two ways. First, because she was always available for sex, even when not ovulating, there was no reason for him to seek other females for sexual pleasure. Second, because her fertility was hidden, he would be motivated to stick around all the time to maximize his own probability of impregnating her and to ensure that no other males mated with her at any time—not just during a brief estrus phase. Fisher says, "Silent ovulation kept a special friend in constant close proximity, providing protection and food the female prized."[18] Known as "mate guarding behavior" to scientists, contemporary women might call it "that insecure pest who never leaves me alone."

Anthropologist Sarah Blaffer Hrdy offers a different explanation for the unusual sexual capacity of the human female. She suggests that concealed ovulation and extended receptivity in early hominids may have evolved not to *reassure* males, but to *confuse* them. Having noted the tendency of newly enthroned alpha male baboons to kill all the babies of the previous patriarch, Hrdy hypothesized that this aspect of female sexuality may have developed as a way of confusing paternity among various males. The female would have sex with several males so that none of them could be certain of paternity, thus reducing the likelihood that the next alpha male would kill offspring who could be his.

So we've got Fisher's "classic theory" proposing that women evolved their special sexiness as a way of keeping one man's interest, and Hrdy saying it's all about keeping several guys guessing. Fisher's theory fits better with the standard model, in which females trade sex for food, protection, and so forth. But this explanation works only if we believe that males—including our "primitive" ancestors—were interested in sex all the time with *just one female*. This contradicts the premise that males are hell-bent on spreading their seed far and wide, while simultaneously protecting their investment in their primary mate/family.

Hrdy's "seeds of confusion" theory posits that concealed ovulation and constant receptivity would benefit a female who had multiple male partners—by preventing them from killing her offspring and inducing them to defend or otherwise aid her children. Hrdy's vision of human sexual evolution puts females directly at odds with males, who would presumably view fertile females as "individually recognizable and potentially defensible resource packets" too valuable to share.

Either way, as depicted in the standard narrative, human sexual prehistory was characterized by deceit, disappointment, and despair. According to this view, both males and females are, by nature, liars, whores, and cheats. At our most basic levels, we're told, heterosexual men and women have evolved to trick one another while selfishly pursuing zero-sum, mutually antagonistic genetic agendas—even though this demands the betrayal of the people we claim to love most sincerely.

Original sin indeed.

CHAPTER FOUR

The Ape in the Mirror

Why should our nastiness be the baggage of an apish past and our kindness uniquely human? Why should we not seek continuity with other animals for our 'noble' traits as well?

STEPHEN JAY GOULD

'Tis from the resemblance of the external actions of animals to those we ourselves perform, that we judge their internal likewise to resemble ours; and the same principle of reasoning, carry'd one step farther, will make us conclude that since our internal actions resemble each other, the causes, from which they are deriv'd, must also be resembling. When any hypothesis, therefore, is advanc'd to explain a mental operation, which is common to men and beasts, we must apply the same hypothesis to both.

DAVID HUME, *A Treatise of Human Nature* (1739–1740)

Genetically, the chimps and bonobos at the zoo are far closer to you and the other paying customers than they are to the gorillas, orangutans, monkeys, or anything else in a cage. Our DNA differs from that of chimps and bonobos by roughly 1.6 percent, making us closer to them than a dog is to a fox, a white-handed gibbon to a white-cheeked crested gibbon, an Indian elephant to an African elephant or, for any bird-watchers who may be tuning in, a red-eyed vireo to a white-eyed vireo.

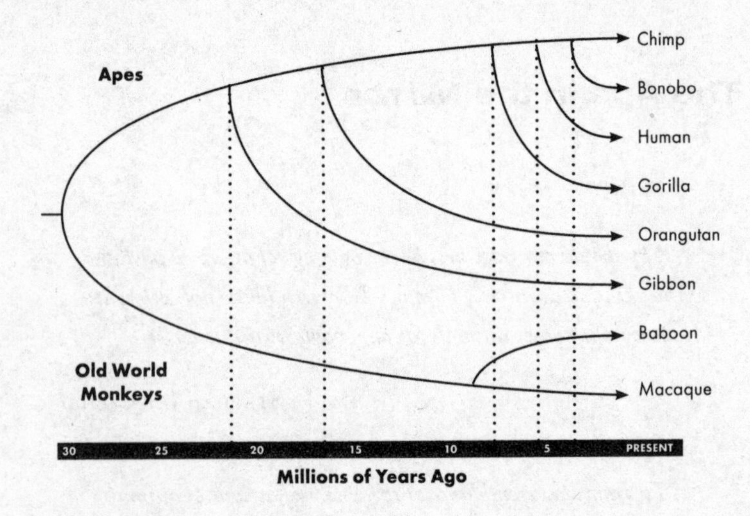

The ancestral line leading to chimps and bonobos splits off from that leading to humans just five to six million years ago (though interbreeding probably continued for a million or so years after the split), with the chimp and bonobo lines separating somewhere between 3 million and 860,000 years ago.[1] Beyond these two close cousins, the familial distances to other primates grow much larger: the gorilla peeled away from the common line around nine million years ago, orangutans 16 million, and gibbons, the only monogamous ape, took an early exit about 22 million years ago. DNA evidence indicates that the last common ancestor for apes and monkeys lived about 30 million years ago. If you picture this relative genetic distance from humans geographically, with a mile representing about 100,000 years since we last shared a common ancestor, it might look something like this:

- *Homo sapiens sapiens*: New York, New York.
- Chimps and bonobos are practically neighbors, living within thirty miles of each other in Bridgeport, Connecticut, and Yorktown Heights, New York. Both just fifty miles from New York, they are well within commuting distance of humanity.
- Gorillas are enjoying cheese-steaks in Philadelphia, Pennsylvania.
- Orangutans are in Baltimore, Maryland, doing whatever it is people do in Baltimore.
- Gibbons are busily legislating monogamy in Washington, D.C.
- Old-world monkeys (baboons, macaques) are down around Roanoke, Virginia.

Carl Linnaeus, the first to make the taxonomic distinction between humans and chimps (in the mid-18th century), came to wish he hadn't. This division (*Pan* and *Homo*) is now regarded as being without scientific justification, and many biologists advocate reclassifying humans, chimps, and bonobos together to reflect our striking similarities.

Nicolaes Tulp, a well-known Dutch anatomist immortalized in Rembrandt's painting *The Anatomy Lesson*, produced the earliest accurate description of a nonhuman ape's anatomy in 1641. The body Tulp dissected so closely resembled a human's that he commented that "it would be hard to find one egg more like another." Although Tulp called his specimen an Indian Satyr, and noted that local people called it an orangutan, contemporary primatologists who have studied Tulp's notes believe it was a bonobo.[2]

Like us, chimps and bonobos are African great apes. Like all apes, they have no tail. They spend a good part of their lives on the ground and are both highly intelligent, intensely social creatures. For bonobos, a turbocharged sexuality utterly divorced from reproduction is a central feature of social interaction and group cohesion. Anthropologist Marvin Harris argues that bonobos get a "reproductive payoff that compensates them for their wasteful approach to hitting the ovulatory target."

The payoff is "a more intense form of social cooperation between males and females" leading to "a more intensely cooperative social group, a more secure milieu for rearing infants, and hence a higher degree of reproductive success for sexier males and females."[3] The bonobo's promiscuity, in other words, confers significant evolutionary benefits on the apes.

The only monogamous ape, the gibbon, lives in Southeast Asia in small family units consisting of a male/female couple and their young—isolated in a territory of thirty to fifty square kilometers. They never leave the trees, have little to no interaction with other gibbon groups, not much advanced intelligence to speak of, and infrequent, reproduction-only copulation.

Monogamy is not found in any social, group-living primate except—if the standard narrative is to be believed—us.

Anthropologist Donald Symons is as amazed as we are at frequent attempts to argue that monogamous gibbons could serve as viable models for human sexuality, writing, "Talk of why (or whether) humans pair bond like gibbons strikes me as belonging to the same realm of discourse as talk of why the sea is boiling hot and whether pigs have wings."[4]

Primates and Human Nature

If Thomas Hobbes had been offered the opportunity to design an animal that embodied his darkest convictions about human nature, he might have come up with something like a chimpanzee. This ape appears to confirm every dire Hobbesian assumption about the inherent nastiness of pre-state existence. Chimps are reported to be power-mad, jealous, quick to violence, devious, and aggressive. Murder, organized warfare between groups, rape, and infanticide are prominent in accounts of their behavior.

Once these chilling observations were published in the 1960s, theorists quickly proposed the "killer ape" theory of human origins. Primatologists Richard Wrangham and Dale Peterson summarize this demonic theory in stark terms, finding in chimpanzee behavior evidence of ancient human blood-lust, writing, "Chimpanzee-like violence preceded and paved the way for human war, making modern humans the dazed survivors of a continuous, 5-million-year habit of lethal aggression."[5]

Before the chimp came to be regarded as the best living model of ancestral human behavior, a much more distant relative, the savanna baboon, held that position. These ground-dwelling primates are adapted to the sort of ecological niche our ancestors likely occupied once they descended from the trees. The baboon model was abandoned when it became clear that they lack some fundamental human characteristics: cooperative hunting, tool use, organized warfare, and power struggles involving complex coalition-building. Meanwhile, Jane Goodall and others were observing these qualities in chimpanzee behavior. Neuroscientist Robert Sapolsky—an expert on baboon behavior—notes that "chimps are what baboons would love to be like if they had a shred of self-discipline."[6]

Perhaps it is not surprising, then, that so many scientists have assumed that chimpanzees are what humans would be like with just a bit *less* self-discipline. The importance of the chimpanzee in late twentieth–century models of human nature cannot be overstated. The maps we devise (or inherit from previous explorers) predetermine where we explore and what we'll find there. The cunning brutality displayed by chimpanzees, combined with the shameful cruelty that characterizes so much of human history, appears to confirm Hobbesian notions of human nature if left unrestrained by some greater force.

Table 1: Social Organization Among Apes[7]

Bonobo	**Egalitarian and peaceful**, bonobo communities are maintained primarily through social **bonding between females**, although females bond with males as well. Male status derives from the

	mother. Bonds between son and mother are lifelong. **Multimale-multifemale mating.**
Chimpanzee	The bonds between males are strongest and lead to constantly shifting **male coalitions**. Females move through overlapping ranges within territory patrolled by males, but don't form strong bonds with other females or any particular male. **Multimale-multifemale mating.**
Human	By far the most diverse social species among the primates, there is plentiful evidence of **all types of socio-sexual bonding, cooperation, and competition** among contemporary humans. **Multimale-multifemale mating.***
Gorilla	Generally, a single dominant male (the so-called "Silverback") occupies a range for his family unit composed of several females and young. Adolescent males are forced out of the group as they reach sexual maturity. Strongest social bonds are between the male and adult females. **Polygynous mating.**
Orangutan	Orangutans are solitary and show little bonding of any kind. Male orangutans do not tolerate each other's presence. An adult male establishes a large territory where several females live. Each has her own range. **Mating is dispersed**, infrequent and often violent.
Gibbon	Gibbons establish nuclear family units; each couple maintains a territory from which other pairs are excluded. **Mating is monogamous.**

* Unless you're sticking with the standard model, in which case humans are classified as monogamous or polygynous, depending on the source.

Doubting the Chimpanzee Model

There are, however, some serious problems with turning to chimpanzee behavior to understand prehistoric human societies. While chimps are extremely hierarchical, groups of human foragers are vehemently egalitarian. Meat sharing is precisely the occasion when chimp hierarchy is most evident, yet a successful hunt triggers the leveling mechanisms most important to human foraging societies. Most primatologists agree about the prominence of power-consciousness in chimpanzees. But it may be premature to generalize from observations made at Gombe, given that observations made at different sites—Taï, on the Ivory Coast of western Africa, for example—suggest some wild chimps handle the sharing of meat in ways more reminiscent of human foragers. Primatologist Craig Stanford found that while the chimps at Gombe are "utterly nepotistic and Machiavellian" about meat distribution, the chimps at Taï share the meat among every individual in the hunting group, whether friend or foe, close relative or relative stranger.[8]

So, while data from the chimps studied by Goodall and others at Gombe appear to support the idea that a ruthless and calculating selfishness is typical of chimpanzee behavior, information from other study sites may contradict or undermine this finding. Given the difficulties inherent in observing chimpanzee behavior in the wild, we should be cautious about generalizing from the limited data we have available on free-ranging chimps. And given their indisputable intelligence and social nature, we should be equally suspicious of data collected from captive chimps, which would appear to be no more generalizable than human prisoner behavior would be to humans.

There are also questions concerning how violent chimps are if left undisturbed in their natural habitat. As we discuss in Chapter 13, several factors could have profoundly altered the chimps' observed behavior. Cultural historian Morris Berman explains that if we "change things such as food supplies, population densities, and the possibilities for spontaneous group formation and dissolution, . . . all hell breaks loose—no less for apes than for humans."[9]

Even if we limit ourselves to the chimpanzee model, the dark self-assurance of modern-day neo-Hobbesian pessimists may be unfounded. Evolutionary biologist Richard Dawkins, for example, might be a bit less certain in his gloomy assessment of human nature: "Be warned that if you wish, as I do, to build a society in which individuals cooperate generously and unselfishly towards a common good, you can expect little help from biological nature. Let us try to *teach* generosity and altruism, because we are born selfish."[10] Maybe, but cooperation runs deep in our species too. Recent findings in comparative primate intelligence have led researchers Vanessa Woods and Brian Hare to wonder whether an impulse toward cooperation might actually be the key to our species-defining intelligence. They write, "Instead of getting a jump start with the most intelligent hominids surviving to produce the next generation, as is often suggested, it may have been the more sociable hominids—because they were better at solving problems together—who achieved a higher level of fitness and allowed selection to favor more sophisticated problem-solving over time."[11] Humans got smart, they hypothesize, because our ancestors learned to cooperate.

Innately selfish or not, the effects of food provisioning and habitat depletion on both wild chimpanzees and human foragers suggest that Dawkins and others who argue that humans are *innately* aggressive, selfish beasts should be careful about citing these chimp data in support of their case. Human groups tend to respond to food surplus and storage with behavior like that observed in chimps: heightened hierarchical social organization, intergroup violence, territorial perimeter defense, and Machiavellian alliances. In other words, humans—like chimps—tend to fight when there's something worth fighting over. But for most of prehistory, there was no food surplus to win or lose and no home base to defend.

In Search of Primate Continuity

Two elements women share with bonobos are that their ovulation is hidden from immediate detection and that they have sex throughout their cycle. But here the similarities end. Where are our genital swellings, and where is the sex at the drop of a hat?

FRANS DE WAAL[12]

Sex was an expression of friendship: in Africa it was like holding hands. . . . It was friendly and fun. There was no coercion. It was offered willingly.

PAUL THEROUX[13]

Whatever one concludes about chimp violence and its relevance to human nature, our other closest primate cousin, the bonobo, offers a fascinating counter-model. Just as the chimpanzee seems to embody the Hobbesian vision of human origins, the bonobo reflects the Rousseauian view. Although best known today as the proponent of the *Noble Savage*, Rousseau's autobiography details a fascination with sexuality that suggests that he would have considered bonobos kindred souls had he known of them. De Waal sums up the difference between these two apes' behavior by saying that "the chimpanzee resolves sexual issues with power; the bonobo resolves power issues with sex."

Though bonobos surpass even chimps in the frequency of their sexual behavior, females of both species engage in multiple mating sessions in quick succession with different males. Among chimpanzees, ovulating females mate, on average, from six to eight times per day, and they are often eager to respond to the mating invitations of any and all males in the group. Describing the behavior of female chimps she monitored, primatologist Anne Pusey notes, "Each, after mating within her natal community, visited the other community while sexually receptive . . . They eagerly approached and mated with males from the new community."[14]

Whatever the truth regarding relations between unprovisioned groups of chimpanzees in the wild, unconscious bias rings out in passages like this one: "In war as in romance, bonobos and chimpanzees appear to be strikingly different. When two bonobo communities meet at a range boundary at Wamba . . . not only is there no lethal aggression as sometimes occurs in chimps, there may be socializing and even sex between females and the enemy community's males."[15]

Enemy? When two groups of intelligent primates get together to socialize and have sex with each other, who would think of these groups as *enemies* or such a meeting as *war*? Note the similar assumptions in this account: "Chimpanzees give a special call that alerts others at a distance to the presence of food. As such, this is food sharing of sorts, *but it need not be interpreted as charitable.* A caller faced with more than enough food will lose nothing by sharing it and may benefit later when another chimpanzee reciprocates [emphasis added]."[16]

Perhaps this seemingly cooperative behavior "need not be interpreted as charitable," but what's the unspoken problem with such an interpretation? Why should we seek to explain away what looks like generosity among nonhuman primates, or other animals in general? Is generosity a uniquely human quality? Passages like these make one wonder why, as Gould asked, scientists are loath to see primate continuity in our positive impulses even as many clearly yearn to locate the roots of our aggression deep in primate past.

Just imagine that we had never heard of chimpanzees or baboons and had known bonobos first. We would at present most likely believe that early hominids lived in female-centered societies, in which sex served important social functions and in which warfare was rare or absent.

FRANS DE WAAL[17]

Because they live only in a remote area of dense jungle in a politically volatile country (Democratic Republic of Congo, formerly Zaire), bonobos were one of the last mammals to be studied in their natural habitat. Although their anatomical differences from common chimps were noted as long ago as 1929, until bonobos' radically different behavior became apparent, they were considered a subgroup of chimpanzee—often called "pygmy chimps."

For bonobos, female status is more important than male hierarchy, but even female rank is flexible and not binding. Bonobos have no formalized rituals of dominance and submission like the status displays common to chimps, gorillas, and other primates. Although status is not completely absent, primatologist Takayoshi Kano, who has collected the most detailed information on bonobo behavior in the wild, prefers to use the term "influential" rather than "high-ranking" when describing female bonobos. He believes that females are respected out of affection, rather than because of rank. Indeed, Frans de Waal wonders whether it's appropriate to discuss hierarchy at all among bonobos, noting, "If there is a female rank order, it is largely based on seniority rather than physical intimidation: older females are generally of higher status than younger ones."[18]

Those looking for evidence of matriarchy in human societies might ponder the fact that among bonobos, female "dominance" doesn't result in the sort of male submission one might expect if it were simply an inversion of the male power structures found among chimps and baboons. The female bonobos use their power differently than male primates. Despite their submissive social role, male bonobos appear to be much better off than male chimps or baboons. As we'll see in later discussions of female-dominated societies, human males also tend to fare pretty well when the women are in charge. While Sapolsky chose to study baboons because of the chronically high stress levels males suffer as a result of their unending struggles for power, de Waal notes that bonobos confront a different sort of existence, saying, "in view of their frequent sexual activity and low aggression, I find it hard to imagine that males of the species have a particularly stressful time."[19]

Crucially, humans and bonobos, but not chimps, appear to share a specific anatomical predilection for peaceful coexistence. Both species have what's called a *repetitive microsatellite* (at gene AVPR1A) important to the release of oxytocin. Sometimes called "nature's ecstasy," oxytocin is important in pro-social feelings like compassion, trust, generosity, love, and yes, eroticism. As anthropologist and author Eric Michael Johnson explains, "It is far more parsimonious that chimpanzees lost this repetitive microsatellite than for both humans and bonobos to independently develop the same mutation."[20]

But there is intense resistance to the notion that relatively low levels of stress and a surfeit of sexual freedom could have characterized the human past. Helen Fisher acknowledges these aspects of bonobo life as well as their many correlates in human behavior, and even makes a sly reference to Morgan's *primal horde*:

> These creatures travel in mixed groups of males, females, and young. . . . Individuals come and go between groups, depending on the food supply, connecting a cohesive community of several dozen animals. Here is a primal horde. . . . Sex is almost a daily pastime. . . . Females copulate during most of their menstrual cycles—a pattern of coitus more similar to women's than any other creature's. . . . Bonobos engage in sex to ease tension, to stimulate sharing during meals, to reduce stress while traveling, and to reaffirm friendships during anxious reunions. "Make love, not war" is clearly a bonobo scheme.[21]

Fisher then asks the obvious question, "Did our ancestors do the same?" She seems to be preparing us for an affirmative answer by noting that bonobos "display many of the sexual habits people exhibit on the streets, in the bars and restaurants, and behind apartment doors in New York, Paris, Moscow, and Hong Kong." "Prior to coitus," she writes, "bonobos often stare deeply into each other's eyes." And Fisher assures her readers that, like human beings, bonobos "walk arm in arm, kiss each other's hands and feet, and embrace with long, deep, tongue-intruding French kisses."[22]

It seems that Fisher, who shares our doubts about other aspects of the standard narrative, is about to reconfigure her arguments concerning the advent of long-term pair bonding and other aspects of human prehistory to better reflect these behaviors shared by bonobos and humans. Given the prominent role of chimpanzee behavior in supporting the standard narrative, how can we *not* include the *equally relevant* bonobo data in our conjectures concerning human prehistory? Remember, *we are genetically equidistant from chimps and bonobos.* And as Fisher notes, human sexual behavior has more in common with bonobos' than with that of any other creature on Earth.

But Fisher balks at acknowledging that the human sexual past could have been like the bonobo present, explaining her last-minute 180-degree turnaround by saying, "Bonobos have sex lives quite different from those of other apes." But this isn't true because humans—whose sexual behavior is so similar to that of bonobos, according to Fisher herself—*are apes.* She continues, "Bonobo heterosexual activities also occur throughout most of the menstrual cycle. And female bonobos resume sexual behavior within a year of parturition." Both these otherwise unique qualities of bonobo sexuality are shared by only one other primate species: *Homo sapiens.* But still, Fisher concludes, "Because pygmy chimps [bonobos] exhibit these *extremes of primate sexuality* and because biochemical data suggest [they] emerged as recently as two million years ago, I do *not* feel they make a suitable model for life as it was among hominids twenty million years ago [emphasis added]."[23]

This passage is bizarre on several levels. After writing at length about how strikingly similar bonobo sexual behavior is to that of human beings, Fisher executes a double backflip to conclude that they *don't* make a suitable model for our ancestors. To make matters even more confusing, she shifts the whole discussion to twenty million years ago as if she'd been talking about the last common ancestor of *all apes* as opposed to that shared by chimps, bonobos, and humans, who diverged from a common ancestor only *five* million years ago. In fact, Fisher wasn't talking about such distant ancestors. *The Anatomy of Love*, the book from which we've been quoting, is a beautifully written popularization of her groundbreaking academic work on the "evolution of serial pair-bonding" in humans (not *all apes*) within the past few million years. Furthermore, note how Fisher refers to the very qualities bonobos share with humans as "extremes of primate sexuality."

Further hints of neo-Victorianism appear in Fisher's description of the transition our ancestors made from the treetops to life on land: "Perhaps our primitive female ancestors living in the trees pursued sex with a variety of males to keep friends. Then, when our forbears were driven onto the grasslands of Africa some four million years ago and pair bonding evolved to raise the young, females turned from open promiscuity to clandestine copulations, reaping the benefits of resources and better or more varied genes as well."[24] Fisher *assumes* the advent of pair bonding four million years ago despite the absence of any supporting evidence. Continuing this circular reasoning, she writes:

> Because bonobos appear to be the smartest of the apes, because they have many physical traits quite similar to people's, and because these chimps copulate with flair and frequency, some anthropologists conjecture that bonobos are much like the African hominoid prototype, our last common tree-dwelling ancestor. Maybe pygmy chimps are living relics of our past. But they certainly manifest some fundamental differences in their sexual behavior. For one thing, bonobos do not form long-term pair-bonds the way humans do. Nor do they raise their young as husband and wife. Males do care for infant siblings, but monogamy is no life for them. Promiscuity is their fare.[25]

Here we have crystalline expression of the Flintstonizing that can distort the thinking of even the most informed theorists on the origins of human sexual behavior. We're confident Dr. Fisher will find that what she calls "fundamental differences" in sexual behavior are not differences at all when she looks at the full breadth of information we cover in following chapters. We'll show that husband/wife marriage and sexual monogamy are *far* from universal human behaviors, as she and others have argued. Simply because bonobos raise doubts about the naturalness of human long-term pair bonding, Fisher and most other authorities conclude that they cannot serve as models for human evolution. They begin by assuming that long-term sexual monogamy forms the nucleus of the one and only natural, eternal human family structure and reason backwards from there. Yucatán be damned!

> *I sometimes try to imagine what would have happened if we'd known the bonobo first and chimpanzee only later or not at all. The discussion about human evolution might not revolve as much around violence, warfare, and male dominance, but rather around sexuality, empathy, caring, and cooperation. What a different intellectual landscape we would occupy!*
>
> FRANS DE WAAL, *Our Inner Ape*

The weakness of the "killer ape theory" of human origins becomes clear in light of what's now known about bonobo behavior. Still, de Waal makes a good case that even without the data that became available in the 1970s, the many flaws in the chimp-fortified Hobbesian view eventually would have emerged. He calls attention to the fact that the theory confuses predation with aggression, assumes that tools originated as weapons, and depicts women as "passive objects of male competition." He calls for a new scenario that "acknowledges and explains the virtual absence of organized warfare among today's human foragers, their egalitarian tendencies, and generosity with information and resources across groups."[26]

By projecting recent post-agricultural preoccupations with female fidelity into their vision of prehistory, many theorists have Flintstonized their way right into a cul-de-sac. Modern man's seemingly instinctive impulse to control women's sexuality is not an intrinsic feature of human nature. It is a response to specific historical socioeconomic conditions—conditions very different from those in which our species evolved. This is key to understanding sexuality in the modern world. De Waal is correct that this hierarchical, aggressive, and territorial behavior is of recent origin for our species. It is, as we'll see, an adaptation to the social world that arose with agriculture.

From our perspective on the far bank, Helen Fisher, Frans de Waal, and a few others seem to have ventured out onto the bridge that crosses over the rushing stream of unfounded assumptions about human sexuality—but they dare not cross it. Their positions seem, to us, to be compromises that strain against the most parsimonious interpretation of data they know as well as anyone. Confronted with the unignorable fact that human beings sure don't *act* like a monogamous species, they make excuses for our "aberrant" (yet perplexingly consistent) behavior. Fisher explains the phenomenon of worldwide marital breakdown by arguing that the pair bond evolved to last only until the infant grows to a child who can keep up with the foraging band without fatherly assistance. For his part, de Waal still argues that the nuclear family is "intrinsically human" and the pair-bond is "the key to the incredible level of cooperation that marks our species." But he then suggestively concludes that "our success as a species is intimately tied to the abandonment of the bonobo lifestyle and to a tighter control over sexual expressions."[27] "Abandonment?" Since it's impossible to abandon what one never had, de Waal would presumably agree that hominid sexuality was, at some point, profoundly similar to that of the relaxed, promiscuous bonobo—although he never says so explicitly. Nor has he ventured to say when or why our ancestors abandoned that way of being.[28]

Table 2: Comparison of Bonobo, Chimp, and Human Socio-sexual Behavior and Infant Development[29]

Human and bonobo females **copulate throughout menstrual cycle**, as well as during **lactation** and **pregnancy**. Female chimps are sexually active only 25–40 percent of their cycle.

Human and bonobo infants develop much more slowly than chimpanzees, beginning to play with others at about 1.5 years, much later than chimps.

Like humans, **female bonobos return to the group immediately after giving birth and copulate within months.** They exhibit little fear of infanticide, which has never been observed in bonobos—captive or free-living.

Bonobos and humans enjoy many different copulatory positions, with ventral-ventral (missionary position) appearing to be preferred by bonobo females and rear-entry by males, while chimps prefer rear-entry almost exclusively.

Bonobos and humans often gaze into each other's eyes when copulating and kiss each other deeply. Chimps do neither.

The **vulva** is located between the legs and oriented toward the **front of the body in humans and bonobos**, rather than oriented toward the rear as in chimps and other primates.

Food sharing is highly associated with sexual activity in humans and bonobos, only moderately so in chimps.

There is a high degree of **variability in potential sexual combinations in humans and bonobos**; homosexual activity is common in both, but rare in chimps.

Genital-genital (G-G) rubbing between female bonobos appears to affirm female bonding, is present in all bonobo populations studied (wild and captive), and is completely absent in chimpanzees. Human data on G-G rubbing are presently unavailable. (Attention: ambitious graduate students!)

While sexual activity in chimps and other primates appears to be primarily reproductive, **bonobos and humans utilize sexuality for social purposes** (tension reduction, bonding, conflict resolution, entertainment, etc.).

PART II

Lust in Paradise (Solitary?)

CHAPTER FIVE

Who Lost What in Paradise?

> *[Man] has imagined a heaven, and has left entirely out of it the supremest of all his delights, the one ecstasy that stands first and foremost in the heart of every individual of his race . . . sexual intercourse! It is as if a lost and perishing person in a roasting desert should be told by a rescuer he might choose and have all longed-for things but one, and he should elect to leave out water!*
>
> MARK TWAIN, *Letters from the Earth*

Turns out, the Garden of Eden wasn't really a garden at all. It was anything but a garden: jungle, forest, wild seashore, open savanna, windblown tundra. *Adam and Eve weren't kicked out of a garden. They were kicked into one.*

Think about it. What's a garden? Land under cultivation. Tended. Arranged. Organized. Intentional. Weeds are pulled or poisoned without mercy; seeds are selected and sown. There's nothing free or spontaneous about such a place. Accidents are unwelcome. But the story says that before their fall from grace, Adam and Eve lived carefree, naked, and innocent—lacking nothing. Their world provided what they needed: food, shelter, and companionship.

But after the Fall, the good times were over. Food, previously the gift of a generous world, now had to be earned through hard work. Women suffered in giving birth. And sexual pleasure—formerly guilt-free—became a source of humiliation and shame. Although the biblical story has it that the first humans were expelled from the garden, the narrative clearly got reversed somewhere along the line. The curse suffered by Adam and Eve centers around the exchange of the arguably low-stress, high-pleasure life of foragers (or bonobos) for the dawn-to-dusk toil of a farmer in his garden. Original sin represents the attempt to explain why on Earth our ancestors ever accepted such a raw deal.[1]

The story of the Fall gives narrative structure to the traumatic transition from the take-it-where-you-find-it hunter-gatherer existence to the arduous struggle of agriculturalists. Contending with insects, rodents, weather, and the reluctant Earth itself, farmers were forced to earn their bread by the sweat of their brow rather than just finding the now-forbidden fruit and eating it hand to mouth, as their ancestors had done forever. No wonder foragers have almost never shown any interest in learning farming techniques from Europeans. As one forager put it, "Why should we plant, when there are so many mongongo nuts in the world?"

Books like this one, concerning human nature, are beacons for trouble. On one hand, everybody's an expert. Being human, we all have opinions about human nature. Such an understanding seems to require little more than a modicum of common sense and some attention to our own incessant cravings and aversions. Simple enough.

But making sense of human nature is anything but simple. Human nature has been landscaped, replanted, weeded, fertilized, fenced off, seeded, and irrigated as intensively as any garden or seaside golf course. Human beings have been under cultivation longer than we've been cultivating anything else. Our cultures domesticate us for obscure purposes, nurturing and encouraging certain aspects of our behavior and tendencies while seeking to eliminate those that might be disruptive. Agriculture, one might say, has involved the domestication of the human being as much as of any plant or other animal.[2]

Our sense of the full range of human nature, like our diet, has been steadily reduced. No matter how nourishing it might be, anything wild gets pulled—though as we'll see, some of the weeds growing in us have roots reaching deep into our shared past. Pull them if you want, but they'll just keep coming back again and again.

What gets cultivated—in soil and minds—is not necessarily beneficial to the individuals in a given society. Something may benefit a culture overall, while being disastrous to the majority of the individual members of that society. Individuals suffer and die in wars from which a society may benefit greatly. Industrial poisons in the air and water, globalized trade accords, genetically modified crops . . . all are accepted by individuals likely to end up losing in the deal.

This disconnect between individual and group interests helps explain why the shift to agriculture is normally spun as a great leap forward, despite the fact that it was actually a disaster for most of the individuals who endured it. Skeletal remains taken from various regions of the world dating to the transition from foraging to farming all tell the same story: increased famine, vitamin deficiency, stunted growth, radical reduction in life span, increased violence . . . little cause for celebration. For most people, we'll see that the shift from foraging to farming was less a giant leap forward than a dizzying fall from grace.

On Getting Funky and Rockin' Round the Clock

If you ever doubt that human beings are, beyond everything, social animals, consider that short of outright execution or physical torture, the worst punishment in any society's arsenal has always been exile. Having run short of empty places to exile our worst prisoners, we've turned to internal exile as our harshest punishment: solitary confinement. Sartre got it backwards when he proclaimed, *"L'enfer, c'est les autres"* (Hell is other people). It's the *absence* of other people that is hellish for our species. Human beings are so desperate for social contact that prisoners almost universally choose the company of murderous lunatics over extended isolation. "I would rather have had the worst companion than no companion at all," said journalist Terry Anderson, recalling his seven-year ordeal as a hostage in Lebanon.[3]

Evolutionary theorists love to seek explanations for species' most outstanding features: the elk's antlers, the giraffe's neck, the cheetah's breakaway speed. These features reflect the environment in which the species evolved and the particular niche it occupies in this environment.

What's our species's outstanding feature? Other than our supersized male genitalia (see Part IV), we're not very impressive from a physical perspective. With less than half our body weight, the average chimp has the strength of any four or five mustachioed firefighters. Plenty of animals can run faster, dive deeper, fight better, see farther, detect fainter smells, and hear tonal subtleties in what sounds like silence to us. So what do we bring to the party? What's so special about human beings?

Our endlessly complex interactions with each other.

We know what you're thinking: big brains. True, but our unique brains result from our chatty sociability. Though debate rages concerning precisely why the human brain grew so large so quickly, most would agree with anthropologist Terrence W. Deacon when he writes, "The human brain has been shaped by evolutionary processes that elaborated the capacities needed for language, and not just by a general demand for greater intelligence."[4]

In a classic feedback loop, our big brains both serve our need for complex, subtle communication *and* result from it. Language, in turn, enables our deepest, most human feature: the ability to form and maintain a flexible, multidimensional, adaptive social network. Before and beyond anything else, human beings are the most social of all creatures.

We have another quality that is *especially human* in addition to our disproportionately large brains and associated capacity for language. Perhaps unsurprisingly, it is also something woven into our all-important social fabric: our exaggerated sexuality.

No animal spends more of its allotted time on Earth fussing over sex than *Homo sapiens*—not even the famously libidinous bonobo. Although we and the bonobo both average well into the hundreds, if not thousands, of acts of intercourse per birth—way ahead of any other primate—their "acts" are far briefer than ours. Pair-bonded "monogamous" animals are almost always *hypo*sexual, having sex as the Vatican recommends: infrequently, quietly, and for reproduction only. Human beings, regardless of religion, are at the other end of the libidinal spectrum: *hyper*sexuality personified.

Human beings and bonobos use eroticism for pleasure, for solidifying friendship, and for cementing a deal (recall that historically, marriage is more akin to a corporate merger than a declaration of eternal love). For these two species (and apparently *only* these two species), nonreproductive sex is "natural," a defining characteristic.[5]

Does all this frivolous sex make our species sound "animalistic"? It shouldn't. The animal world is full of species that have sex only during widely spaced intervals when the female is ovulating. Only two species can do it week in and week out for nonreproductive reasons: one human, the other very humanlike. Sex for pleasure with various partners is therefore more "human" than animal. Strictly reproductive, once-in-a-blue-moon sex is more "animal" than human. In other words, an excessively horny monkey is acting "human," while a man or woman uninterested in sex more than once or twice a year would be, strictly speaking, "acting like an animal."

Though many strive to hide their human libidinousness from themselves and each other, being a force of nature, it breaks through. Lots of upright, proper Americans were scandalized by the way Elvis moved his hips when he sang "rock and roll." But how many realized what the phrase *rock and roll* meant? Cultural historian Michael Ventura, investigating the roots of African-American music, found that rock 'n' roll was a term that originated in the juke joints of the South. Long in use by the time Elvis appeared, Ventura explains the phrase "hadn't meant the name of a music, it meant 'to fuck.' 'Rock,' by itself, had pretty much meant that, in those circles, since the twenties at least." By the mid-1950s, when the phrase was becoming widely used in mainstream culture, Ventura says the disc jockeys "either didn't know what they were saying or were too sly to admit what they knew."

Though crusty old Ed Sullivan would have been scandalized to realize what he was saying when he announced this new "rock and roll all the kids are crazy about," examples of barely concealed sexual reference lurking just below the surface of common American English don't stop there. Robert Farris Thompson, America's most prominent historian of African art, says that *funky* is derived from the Ki-Kongo *lu-fuki*, meaning "positive sweat" of the sort you get from dancing or having sex, but not working. One's *mojo*, which has to be "working" to attract a lover, is Ki-Kongo for "soul." *Boogie* comes from *mbugi*, meaning "devilishly good." And both *jazz* and *jism* likely derive from *dinza*, the Ki-Kongo word for "to ejaculate."[6]

Forget the billions pouring in from porn. Forget all the T&A on TV, in advertising, and in movies. Forget the love songs we sing on the way into relationships and the blues on the way out. Even if we include none of that, the percentage of our lives we human beings spend thinking about, planning, having, and remembering sex is incomparably greater than that of any other creature on the planet. Despite our relatively low reproductive potential (few women have ever had more than a dozen or so children), our species truly can, and does, rock around the clock.

If I had had to choose my place of birth, I would have chosen a state in which everyone knew everyone else, so that neither the obscure tactics of vice nor the modesty of virtue could have escaped public scrutiny and judgment.

JEAN-JACQUES ROUSSEAU,
"Discourse on the Origin of Inequality" (1754)

Rousseau was born in the wrong time, wrong place. If he'd been born in the same spot twenty thousand years earlier, among the artists sketching life-sized bulls on European cave walls, he'd have known every member of his social world. Alternatively, born into his own era but in one of the many societies not yet altered by agriculture, he'd have found the close-knit social world for which he yearned. The sense of being alone—even in a crowded city—is an oddity in human life, included, like so much else, in the agricultural package.

Looking back from his overcrowded world, Thomas Hobbes imagined that prehistoric human life was unbearably solitary. Today, separated from countless strangers by only thin walls, tiny earphones, and hectic schedules, we assume a desolate sense of isolation must have weighed on our ancestors, wandering over their windswept prehistoric landscape. But in fact, this seemingly common-sense assumption couldn't be more mistaken.

The social lives of foragers are characterized by a depth and intensity of interaction few of us could imagine (or tolerate). For those of us born and raised in societies organized around the interlocking principles of individuality, personal space, and private property, it's difficult to project our imaginations into those tightly woven societies where almost all space and property is communal, and identity is more collective than individual. From the first morning of birth to the final mourning of death, a forager's life is one of intense, constant interaction, interrelation, and interdependence.

In this section, we'll examine the first element in Hobbes's famous dictum about prehistoric human life. We'll show that before the rise of the state, prehistoric human life was far from "solitary."

Societies Mentioned in Text

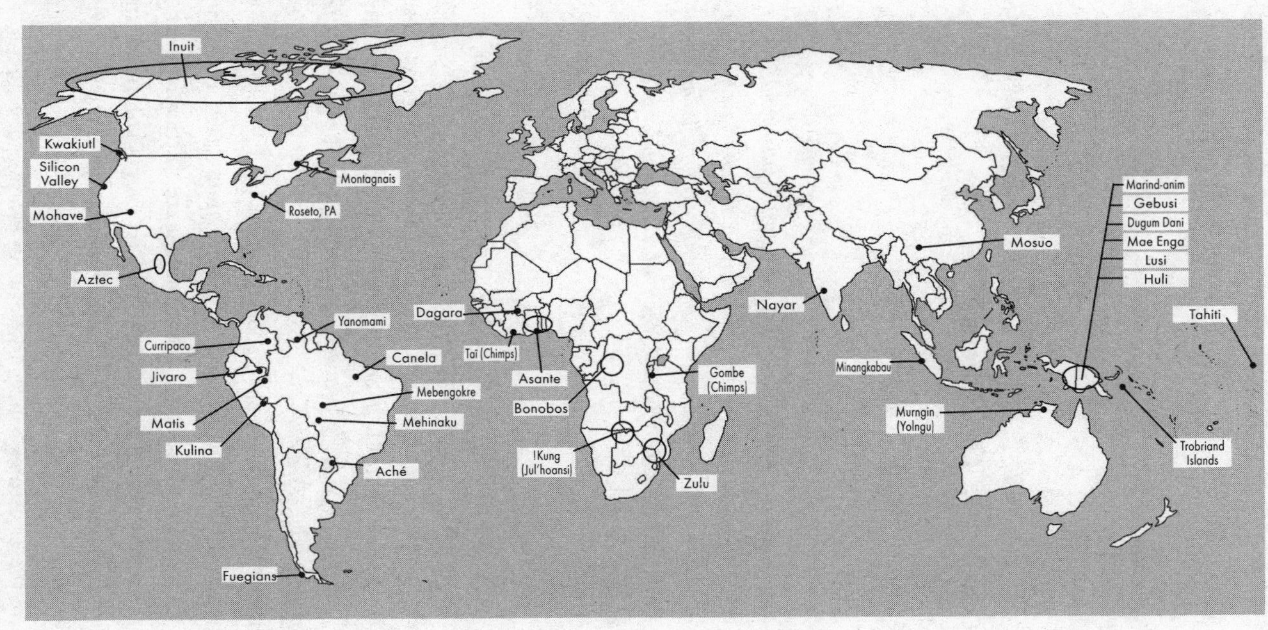

CHAPTER SIX

Who's Your Daddies?

> *In view of the frequent occurrence of modern domestic groups that do not consist of, or contain, an exclusive pair-bonded father and mother, I cannot see why anyone should insist that our ancestors were reared in monogamous nuclear families and that pair-bonding is more natural than other arrangements.*
>
> MARVIN HARRIS[1]

The birds and the bees are different in the Amazon. There, a woman not only *can* be a little pregnant, most are. Each of the societies we're about to discuss shares a belief in what scientists call "partible paternity." These groups have a novel conception of conception: *a fetus is made of accumulated semen.*

Anthropologists Stephen Beckerman and Paul Valentine explain, "Pregnancy is viewed as a matter of degree, not clearly distinguished from gestation . . . all sexually active women are a little pregnant. Over time . . . semen accumulates in the womb, a fetus is formed, further acts of intercourse follow, and additional semen causes the fetus to grow more."[2] Were a woman to stop having sex when her periods stopped, people in these cultures believe the fetus would stop developing.

This understanding of how semen forms a child leads to some mighty interesting conclusions regarding "responsible" sexual behavior. Like mothers everywhere, a woman from these societies is eager to give her child every possible advantage in life. To this end, she'll typically seek out sex with an assortment of men. She'll solicit "contributions" from the best hunters, the best storytellers, the funniest, the kindest, the best-looking, the strongest, and so on—in the hopes her child will literally absorb the essence of each.

Anthropologists report similar understandings of conception and fetal development among many South American societies, ranging from simple hunter-gatherers to horticulturalists. A partial list would include the Aché, the Araweté, the Barí, the Canela, the Cashinahua, the Curripaco, the Ese Eja, the Kayapó, the Kulina, the Matis, the Mehinaku, the Piaroa, the Pirahã, the Secoya, the Siona, the Warao, the Yanomami, and the Ye'kwana—societies from Venezuela to Bolivia. This is no ethnographic curiosity, either—a strange idea being passed among related cultures. The same understanding is found among cultural groups that show no evidence of contact for millennia. Nor is partible paternity limited to South America. For example, the Lusi of Papua New Guinea also hold that fetal development depends on multiple acts of intercourse, often with different men. Even today, the younger Lusi, who have some sense of the modern understanding of reproduction, agree that a person can have more than one father.

As Beckerman and Valentine explain, "It is difficult to come to any conclusion except that partible paternity is an ancient folk belief capable of supporting effective families, families that provide satisfactory paternal care of children and manage the successful rearing of children to adulthood."[3]

When an anthropologist working in Paraguay asked his Aché subjects to identify their fathers, he was presented with a mathematical puzzle that could be solved only with a vocabulary lesson. The 321 Aché claimed to have over six hundred fathers. Who's your daddies?

It turns out the Aché distinguish four different kinds of fathers. According to the anthropologist Kim Hill, the four types of fathers are:

- *Miare*: the father who put it in;
- *Peroare*: the fathers who mixed it;
- *Momboare*: those who spilled it out; and
- *Bykuare*: the fathers who provided the child's essence.[4]

Rather than being shunned as "bastards" or "sons of bitches," children of multiple fathers benefit from having more than one man who takes a special interest in them. Anthropologists have calculated that their chances of surviving childhood are often significantly better than those of children in the same societies with just one recognized father.[5]

Far from being enraged at having his genetic legacy called into question, a man in these societies is likely to feel *gratitude* to other men for pitching in to help create and then care for a stronger baby. Far from being blinded by jealousy as the standard narrative predicts, men in these societies find themselves bound to one another by shared paternity for the children they've fathered together. As Beckerman explains, in the worst-case scenario, this system may provide extra security for the child: "You know that if you die, there's some other man who has a residual obligation to care for at least one of your children. So looking the other way or even giving your blessing when your wife takes a lover is the only insurance you can buy."[6]

Lest any readers feel tempted to file this sort of behavior under B.A.D. (Bizarre And Distant), similar examples can be found quite close to home.

The Joy of S.E.Ex.

Understanding is a lot like sex; it's got a practical purpose, but that's not why people do it normally.

FRANK OPPENHEIMER

Desmond Morris spent months observing a British pro soccer team in the late 1970s and early 1980s, later publishing his thoughts in a book called *The Soccer Tribe.* As his title suggests, Morris found the behavior of the teammates to be strikingly similar to what he'd encountered among tribal groups in previous research. He noted two behaviors particularly salient in both contexts: group leveling and nonpossessiveness.

"The first thing you notice when footballers talk among themselves," Morris wrote, "is the speed of their wit. Their humour is often cruel and is used to deflate any team-mate who shows the slightest signs of egotism." But echoes of prehistoric egalitarianism reverberate beyond ego deflation in the locker room, extending to sexuality as well. "If one of them scores (sexually), he is not possessive, but is only too happy to see his team-mates succeed with the same girl." While this may strike some as unfeeling, Morris assured his readers that this lack of jealousy was "simply a measure of the extent to which selfishness is suppressed between team-mates, both on the field and off it."[7]

For professional athletes, musicians, and their most enthusiastic female fans, as well as both male and female members of many foraging societies, overlapping, intersecting sexual relationships strengthen group cohesion and can offer a measure of security in an uncertain world. Sometimes, perhaps most of the time, human sex isn't just about pleasure or reproduction. A casual approach to sexual relationships in a community of adults can have important social functions, extending far beyond mere physical gratification.

Let's try putting this liquid libido into dry, academic terms: we hypothesize that *Socio-Erotic Exchanges* (S.E.Ex. for short) strengthen the bonds among individuals in small-scale nomadic societies (and, apparently, other highly interdependent groups), forming a crucial, durable web of affection, affiliation, and mutual obligation.

In evolutionary terms, it would be hard to overstate the importance of such networks. After all, it was primarily such flexible, adaptive social groups (and the feedback loop of brain growth and language capacities that both allowed and resulted from them) that enabled our slow, weak, generally unimpressive species to survive and eventually dominate the entire planet. Without frequent S.E.Ex., it's doubtful that foraging bands could have maintained social equilibrium and fecundity over the millennia. S.E.Ex. were crucial in binding adults into groups that cared communally for children of obscure or shared paternity, each child likely related to most or all of the men in the group (if not a father, certainly an uncle, cousin . . .).*

Because these interlocking relationships are so crucial to social cohesion, opting out can cause problems. Writing of the Matis people, anthropologist Philippe Erikson confirms, "Plural paternity . . . is more than a theoretical possibility. . . . Extramarital sex is not only widely practiced and usually tolerated, in many respects, it also appears *mandatory*. Married or not, one has a *moral duty* to respond to the sexual advances of opposite-sex cross-cousins (real or classificatory), under pains of being labeled 'stingy of one's genitals,' a breach of Matis ethics far more serious than plain infidelity [emphasis added]."[8]

Being labeled a sexual cheapskate is no laughing matter, apparently. Erikson writes of one young man who cowered in the anthropologist's hut for hours, hiding from his horny cousin, whose advances he couldn't legitimately reject if she tracked him down. Even more serious, during Matis tattooing festivals, having sex with one's cus-

---

* In a personal communication, Don Pollock makes an interesting point about the notion of multiple fatherhood, writing, "I have always found the Kulina notion that more than one man may be a 'biological' father to a child to be, ironically, similar to the genetic reality: in a small, genetically homogeneous population (or close to it, after many generations of in-marrying), every child has close genetic similarities to all of the men with whom his or her mother had sexual relations—even to those the mother was not involved with."

tomary partner(s) is expressly forbidden—under threat of extreme punishment, even death.[9]

But if it's true that S.E.Ex. played a central role in maintaining prehistoric social cohesion, we should find remnants of such shamelessly libidinous behavior throughout the world, past and present. We do.

Among the Mohave, women were famous for their licentious habits and disinclination to stick with one man.[10] Caesar (yes, *that* Caesar) was scandalized to note that in Iron Age Britain, "Ten and even twelve have wives common to them, and particularly brothers among brothers. . . ."[11] During his three months in Tahiti in 1769, Captain James Cook and his crew found that Tahitians "gratified every appetite and passion before witnesses." In an account of Cook's voyage first published in 1773, John Hawkesworth wrote of "[a] young man, nearly six feet high, perform[ing] the rites of Venus with a little girl about 11 or 12 years of age, before several of our people and a great number of natives, without the least sense of its being indecent or improper, but, as appeared, in perfect conformity to the custom of the place." Some of the older islander women who were observing this amorous display apparently called out instructions to the girl, although Cook tells us, "Young as she was, she did not seem much to stand in need of [them]."[12]

Samuel Wallis, another ship captain who spent time in Tahiti, reported, "The women in General are very handsome, some really great Beauties, yet their Virtue was not proof against a Nail." The Tahitians' fascination with iron resulted in a de-facto exchange of a single nail for a sexual tryst with a local woman. By the time Wallis set sail, most of his men were sleeping on deck, as there were no nails left from which to hang their hammocks.[13]

There is a yam-harvest festival in the present-day Trobriand Islands, in which groups of young women roam the islands "raping" men from outside their own village, purportedly biting off their eyebrows if the men do not satisfy them. Ancient Greece celebrated sexual license in the festivals of Aphrodisia, Dionysia, and Lenea. In Rome, members of the cult of Bacchus hosted orgies no fewer than five times per month, while many islands in the South Pacific are still famous for their openness to unconstrained sexuality, despite the concerted efforts of generations of missionaries preaching the morality of shame.[14] Many modern-day Brazilians let it all hang out during Carnival, when they participate in a rite of consensual nonmarital sex known as *sacanagem* that makes the goings-on in New Orleans or Las Vegas look tame.

Though the eager participation of women in these activities may surprise some readers, it has long been clear that the sources of female sexual reticence are more cultural than biological, despite what Darwin and others have supposed. Over fifty years ago, sex researchers Clellan Ford and Frank Beach declared, "In those societies which have no double standard in sexual matters and in which a variety of liaisons are permitted, the women avail themselves as eagerly of their opportunity as do the men."[15]

Nor do the females of our closest primate cousins offer much reason to believe the human female *should* be sexually reluctant due to purely biological concerns. Instead, primatologist Meredith Small has noted that female primates are highly attracted to novelty in mating. Unfamiliar males appear to attract females more than known males with *any* other characteristic a male might offer (high status, large size, coloration, frequent grooming, hairy chest, gold chains, pinky ring, whatever). Small writes, "The only consistent interest seen among the general primate population is an interest in novelty and variety. . . . In fact," she reports, "the search for the unfamiliar is documented as a female preference more often than is any other characteristic our human eyes can perceive."[16]

Frans de Waal could have been referring to any of the previously mentioned Amazonian societies when he wrote that the male "has no idea which copulations may result in conception and which may not. Almost any [child] growing up in the group could be his. . . . If one had to design a social system in which fatherhood remained obscure, one could hardly do a better job than Mother Nature did with [this] society."[17] Though de Waal's words are applicable to any of the many societies who engage in ritualized extra-pair sex, he was, in fact, writing of the bonobo, thus underscoring the sexual continuity linking the three most closely related apes: chimps, bonobos, and their conflicted human cousins.

In light of the hypersexuality of humans, chimps, and bonobos, one wonders why so many insist that female sexual exclusivity has been an integral part of human evolutionary development for over a million years. In addition to all the direct evidence presented here, the circumstantial case against the narrative is overwhelming.

For starters, recall that the total number of monogamous primate species that live in large social groups is precisely *zero*—unless you insist on counting humans as the one and only example of such a beast. The few monogamous primates that do exist (out of hundreds of species) all live in the treetops. Primates aside, only 3 percent of mammals and one in ten thousand invertebrate species can be considered sexually monogamous. Adultery has been documented in *every* ostensibly monogamous human society ever studied, and is a leading cause of divorce all over the world today. But even in the latest editions of his classic book *The Naked Ape*, the same Desmond Morris who observed soccer players happily sharing their lovers still insists that "among humans sexual behavior occurs almost exclusively in a pair-bonded state," and that "adultery reflects an imperfection in the pair-bonding mechanism."[18]

That's a major minor "imperfection."

As we write these words, CNN reports that six adulterers are being stoned to death in Iran. Before the hypocritical sinners throw the first stones, the male adulterers will be buried up to their waists. In a sickening gesture toward chivalry, the women will be buried to their necks, presumably to bring a quicker death to these women who dared consider their bodies their own. Such brutal execution of sexual transgressors is anything but an oddity, historically speaking. "Judaism, Christianity, Islam and Hinduism each share a fundamental concern over the punishment for a woman's sexual freedom," says Eric Michael

Johnson. "Whereas any 'man that committeth adultery with another man's wife [both] the adulterer and the adulteress shall be put to death,' (Leviticus 20:10) but any unmarried woman who has sexual relations with an unmarried man shall be brought 'to the door of her father's house, and the men of her city shall stone her with stones that she die' (Deuteronomy 22:21)."[19]

Yet even after centuries of such barbaric punishment, adultery persists everywhere, without exception. As Alfred Kinsey noted back in the 1950s, "Even in cultures which most rigorously attempt to control the female's extramarital coitus, it is perfectly clear that such activity does occur, and in many instances it occurs with considerable regularity."[20]

Think about that. *No* group-living nonhuman primate is monogamous, and adultery has been documented in *every* human culture studied—including those in which fornicators are routinely stoned to death. In light of all this bloody retribution, it's hard to see how monogamy comes "naturally" to our species. Why would so many risk their reputations, families, careers—even presidential legacies—for something that runs *against* human nature? Were monogamy an ancient, evolved trait characteristic of our species, as the standard narrative insists, these ubiquitous transgressions would be infrequent and such horrible enforcement unnecessary.

No creature needs to be threatened with death to act in accord with its own nature.

The Promise of Promiscuity

Modern men and women are obsessed with the sexual; it is the only realm of primordial adventure still left to most of us. Like apes in a zoo, we spend our energies on the one field of play remaining; human lives otherwise are pretty well caged in by the walls, bars, chains, and locked gates of our industrial culture.

EDWARD ABBEY

As we consider alternate views of prehistoric human sexuality, keep in mind that the core logic of the standard narrative pivots on two interlocking assumptions:

- A prehistoric mother and child needed the meat and protection a man would provide.
- A woman would have had to offer her own sexual autonomy in exchange, thus assuring him that it was his child he was supporting.

The standard narrative is founded upon the belief that the exchange of protein and protection for assured paternity was the best way to increase the odds of a child's survival to reproductive age. Survival of offspring is, after all, the primary engine of natural selection as described by Darwin and subsequent theorists. But what if risk to offspring were mitigated more effectively by behavior that encouraged the *opposite* arrangement? What if, rather than one man agreeing to share his meat, protection, and status with a particular woman and her child, sharing were generalized? What if group-wide sharing offered a more effective approach to the risks our ancestors encountered in the prehistoric world? And in light of these risks, what if paternity *uncertainty* were more beneficial to the child's chances of survival, as more men would take an interest in him or her?

Again, we're not suggesting a *nobler* social system, just one that might have been better suited to meeting the challenges of prehistoric conditions and more effective in helping people survive long enough to reproduce.

This sharing-based social life is far from uniquely human. For example, vampire bats in Central America feed on the blood of large mammals. But not every bat finds a meal each night. When they return to their dens, those bats who have had a good night regurgitate blood into the mouths of bats who have not had as much luck. Recipients of such largess are likely to return the favor when the conditions have reversed themselves, but are less likely to give blood to bats who have denied them in the past. As one reviewer put it, "The key to this bit for bat process is the individual bat's ability to remember the history of its relationships with all other bats living in its den. This mnemonic requirement has driven the evolution of vampire bat brains, which possess the largest neocortex of all known bat species."[21]

We hope the thought of vampire bats coughing up (coughing down?) blood for their non–blood relatives is graphic enough to convince you that sharing isn't innately "noble." Some species, in some conditions, have simply found that generosity is the best way to reduce risk in an uncertain ecological context. *Homo sapiens* appears to have been such a species until relatively recent times.[22]

The near universality of fierce egalitarianism among foragers suggests there was really little choice for our prehistoric ancestors. Archaeologist Peter Bogucki writes, "For Ice Age mobile hunting societies, the band model of social organization, with its obligatory sharing of resources, was really the only way to live."[23] It makes perfect Darwinian sense to suppose that prehistoric humans would choose the path that offered the best chance of survival—even if that path required egalitarian sharing of resources rather than the self-interested hoarding of resources many contemporary Western societies insist is basic human nature. After all, Darwin himself believed a tribe of cooperative people would vanquish one composed of selfish individualists.

Are we preaching far-fetched flower-power silliness? Hardly. Egalitarianism is found in nearly all simple hunter-gatherer societies that have been studied anywhere in the world—groups facing conditions most similar to those our ancestors confronted 50,000 or 100,000 years ago. They've followed an egalitarian path not because they are particularly noble, but *because it offers them the best chance of survival.* Indeed, under these conditions, egalitarianism may be the *only* way to live, as Bogucki concluded. Institutionalized sharing of resources and sexuality spreads and minimizes risk, assures food won't be wasted in a world without refrigeration, eliminates the effects of male infertility, promotes the genetic health of individuals, and assures a more secure social environment for children and adults alike. Far from utopian romanticism, foragers insist on egalitarianism because it works on the most practical levels.

Bonobo Beginnings

The effectiveness of sexual egalitarianism is confirmed by female bonobos, who share many otherwise unique traits with humans and no other species. These sexual characteristics have direct, predictable social consequences. De Waal's research has demonstrated, for example, that the increased sexual receptivity of the female bonobo dramatically reduces male conflict, when compared with other primates whose females are significantly less sexually available. The abundance of sexual opportunity makes it less worthwhile for males to risk injury by fighting over any particular sexual opportunity. Since alliances among male chimps, for example, generally serve to keep competitors away from an ovulating female, or to attain the high status that brings more mating opportunities to a given male, the principal motivation for these unruly gangs evaporates in the relaxing heat of bonobos' plentiful sexual opportunity.

These same dynamics apply to human groups. Aside from "the social habits of man as he now exists," why presume the monogamous pair-based model of human evolution currently favored would have been adaptive for early humans, but not for bonobos in the jungles of central Africa? Unconstrained by cultural restrictions, the so-called *continual responsiveness* of the human female would fulfill the same function: provide plentiful sexual opportunity for males, thereby reducing conflict and allowing larger group sizes, more extensive cooperation, and greater security for all. As anthropologist Chris Knight puts it, "Whereas the basic primate pattern is to deliver a periodic 'yes' signal against a background of continuous sexual 'no', humans [and bonobos] emit a periodic 'no' signal against a background of continuous 'yes'."[24]

Here we have the *same behavioral and physiological adaptation*, unique to two very closely related primates, yet many theorists insist the adaptation must have completely different origins and functions in each.

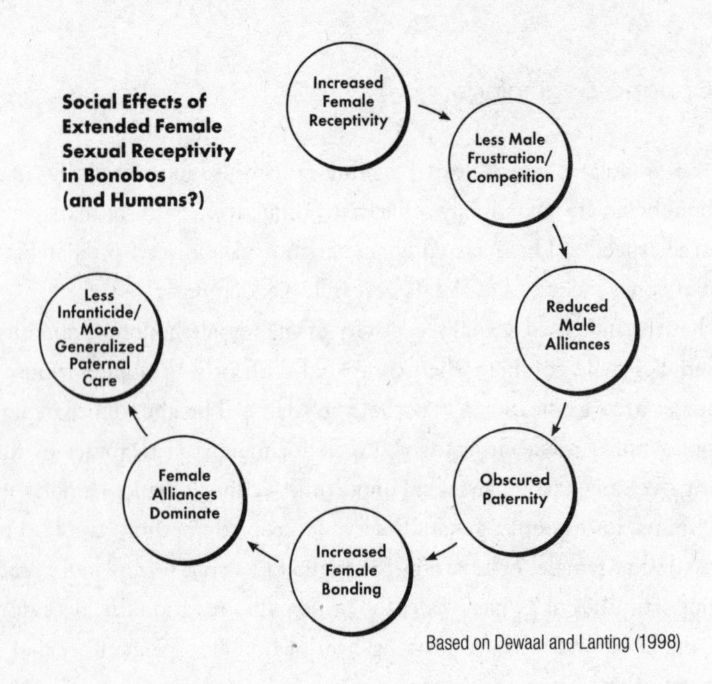

Based on Dewaal and Lanting (1998)

This increased social cohesion is, in fact, probably the most common explanation for the potent combination of extended receptivity and hidden ovulation found *only* in humans and bonobos.[25] But most scientists seem to see only half of this logical connection, as in this abstract: "Females who concealed ovulation were favored because the group in which they lived maintained a peaceful stability that facilitated monogamy, sharing and cooperation."[26] It's clear how greater female sexual availability could increase sharing, cooperation, and peaceful stability, but why monogamy should be added to the list is a question that not only goes unanswered but is almost never asked.

Those anthropologists willing to acknowledge the realities of human sexuality see its social benefits clearly. Beckerman and Valentine point to the fact that partible paternity defuses potential conflicts between men, noting that such antagonisms tend to be unhelpful to a woman's long-term reproductive interests. Anthropologist Thomas Gregor reported eighty-eight ongoing affairs among the thirty-seven adults in the Mehinaku village he studied in Brazil. In his opinion, extramarital relationships "contribute to village cohesion," by "consolidating relationships between persons of different [clans]" and "promoting enduring relationships based on mutual affection." He found that "many lovers are very fond of one another and regard separation as a privation to avoid."[27]

Rather than risk overwhelming you with dozens more examples of this community-building, conflict-reducing human sexuality, we'll conclude with just one more. Anthropologists William and Jean Crocker visited and studied the Canela people—also of the Brazilian Amazon region—for more than three decades, beginning in the late 1950s. They explain:

> It is difficult for members of a modern individualistic society to imagine the extent to which the Canela saw the group and the tribe as more important than the individual. Generosity and sharing was the ideal, while withholding was a social evil. Sharing possessions brought esteem. Sharing one's body was a direct corollary. Desiring control over one's goods and self was a form of stinginess. In this context, it is easy to understand why women chose to please men and why men chose to please women who expressed strong sexual needs. *No one was so self-important that satisfying a fellow tribesman was less gratifying than personal gain* [emphasis in original].[28]

Recognized as a way to build and maintain a network of mutually beneficial relationships, nonreproductive sex no longer requires special explanations. Homosexuality, for example, becomes far less confusing, in that it is, as E. O. Wilson has written, "above all a form of bonding . . . consistent with the greater part of heterosexual behavior as a device that cements relationships."[29]

Paternity certainty, far from being the universal and overriding obsession of all men everywhere and always, as the standard narrative insists, was likely a nonissue to men who lived before agriculture and resulting concerns with passing property through lines of paternal descent.

CHAPTER SEVEN

Mommies Dearest

The diffused sense of parental responsibility resulting from these intersecting webs of sexual interaction extends to mothers as well as fathers. Anthropologist Donald Pollock tells us that while the Kulina believe the fetus to have originally been formed of accumulated semen (*men's milk*, in Kulina), they attribute the baby's growth after birth to *women's milk*. "Any number of women might nurse the child," he writes. "It is particularly common for a group of sisters . . . to share nursing functions; it is not unknown for the mother's mother to allow an infant to nurse, even if the grandmother is no longer lactating, to quiet a crying child whose mother is occupied." When he asked whether these other women were also mothers of the child, Pollock was told this was "obviously so."[1]

Recalling his childhood among the Dagara, in Burkina Faso, author and psychologist Malidoma Patrice Somé remembers how freely children wandered into houses throughout the village. Somé explains that this "gives the child a very broad sense of belonging," and that "everybody chips in to help raise the child." Apart from the many obvious benefits to parents, Somé sees distinct psychological advantages for the children, saying, "It's very rare that a child feels isolated or develops psychological problems; everyone is very aware of where he or she belongs."[2]

Though Somé's account may sound like idealized memory, what he describes is still standard village life in most of rural Africa, where children are welcome to wander in and out of the homes of unrelated adults in villages. Though a mother's love is no doubt unique, women (and some men) the world over are eager to coo over unrelated babies, not just their own—an eagerness common to other social primates, none of whom, by the way, are monogamous. This deeply felt, broadly shared willingness to care for unrelated children lives on in the modern world: the bureaucratic ordeal of adoption rivals or exceeds the stress and expense of childbirth, yet millions of couples eagerly pursue its uncertain rewards.

Scientists focused only on the nuclear family miss the central role of alloparenting in our species.* Sarah Blaffer Hrdy, author of *Mothers and Others*, laments, "Infant-sharing in other primates and in various tribal societies has never been accorded center stage in the anthropological literature. Many people don't even realize it goes on. Yet . . . the consequences of cooperative care—in terms of survival and biological fitness of mother and infant—turn out to be all to the good."[3]

Darwin entertained the radical possibility that the mother-child bond may have been less important to "barbarous" individuals than their bond with the greater group. Commenting on the customary use of familial terms like *mother*, *father*, *son*, and *daughter* in reference to all group members, he suggested, "The terms employed express a connection with the tribe alone, to the exclusion of the mother. It seems possible that the connection between the related members of the same barbarous tribe, exposed to all sorts of danger, might be so much more important, owing to the need of mutual protection and aid, than that between the mother and her child. . . ."[4]

When seventeenth-century Jesuit missionary Paul Le Jeune lectured a Montagnais Indian man about the dangers of the rampant infidelity he'd witnessed, Le Jeune received a lesson on proper parenthood in

* Where nonparents act in a parental role.

response. The missionary recalled, "I told him that it was not honorable for a woman to love any one else except her husband, and that this evil being among them, he himself was not sure that his son, who was there present, was his son. He replied, 'Thou hast no sense. You French people love only your own children; but we all love all the children of our tribe.'"[5]

Though it seems like common sense to most of us, our own biology-based kinship system is another case of Flintstonization. We simply *assume* our own conception of family reflects something eternal and universal in human nature. But as we've seen, there isn't even agreement among all people that one sex act is sufficient to result in pregnancy.

The concept of one-mother-per-child is running into trouble in Western societies too. "Motherhood is splintering," writes William Saletan, the "human nature expert" at Slate.com. "You can have a genetic mother, a gestational mother, an adoptive mother, and God knows what else. When one of your moms is Grandma, it's even more confusing." Speaking of surrogate mothers who gestate another woman's fetus, Saletan argues that it makes sense for a woman's mother to offer to carry the baby: "When the surrogate is Grandma, the mess is less. Mother and daughter share a genetic bond to each other and to the child. They're much more likely to work things out and give the child a stable family environment."[6] Perhaps. Either way, with widespread adoption, stepfamilies resulting from remarriage, and techniques such as surrogate gestation, sperm donation, and cryogenic embryo preservation, *Homo sapiens* is on the fast track away from "traditional" family structures, perhaps headed toward more flexible arrangements reminiscent of the distant past.

Belief in partible paternity spreads fatherly feelings throughout a group, but this is just one of many mechanisms for enhancing group solidity. Anthropologists report numerous societies in which naming ceremonies and clan affiliations create obligations between individuals more binding than blood relations. Referring to the Matis people, with whom he lived, anthropologist Philippe Erikson notes, "When it comes to defining kinship ties, relationships deriving from naming practices have absolute priority over any other considerations, such as genealogical connections. When conflicts arise between the two modes of reckoning, sharing a name has precedence. . . ."[7]

Some anthropologists question whether kinship is an important concept in band-level societies at all—however defined. They argue that since everyone in such a small-scale society is likely to be related to each other in some way, affinity tends to be measured in more fluid terms, such as friendship and sharing partners.

As Darwin understood, even the most direct and immediate kinship terminology is subject to cultural definition. "Fatherly behavior is expected of all the males of a local clan toward all the young of the clan," says anthropologist Janet Chernela. "Multiple aspects of caretaking, including affection and food-getting, are provided by all clansmen."[8] Anthropologist Vanessa Lea notes that, based upon her experience among the Mebengokre, "The allocation of responsibility is socially constructed and not an objective fact. . . ."[9] Among the Tukanoan, "Clan brothers provide for one another's children as a collective. Through the pooling of the daily catch, each male regularly labors for all of the children of a village—his own offspring as well as those of his brothers."[10]

This diffused approach to parenting isn't limited to villages in Africa or Amazonia. Desmond Morris recalls an afternoon he spent with a female truck driver in Polynesia. She told him that she'd had nine children, but had given two of them to an infertile friend. When Morris asked how the kids felt about that, she said they didn't mind at all, as "all of us love all of the children." Morris recalls, "This last point is underlined by the fact that, when we reach the village . . . she passes the time by wandering over to a group of toddlers, lying down in the grass with them and playing with them exactly as if they were her own. They accept her instantly, without any questioning, and a passer-by would never have guessed that they were anything other than a natural family playing together."[11]

"A natural family." Perhaps this easy acceptance between adults and unrelated children, the diffuse nurturing found in societies where children refer to all men as *father* and all women as *mother*, societies small and isolated enough to safely assume the kindness of strangers, where overlapping sexual relationships leave genetic paternity unknowable and of little consequence . . . perhaps this *is* the "natural" family structure of our species.

Could it be that the atomic isolation of the husband-wife nucleus with an orbiting child or two is in fact a culturally imposed aberration for our species—as ill-suited to our evolved tendencies as corsets, chastity belts, and suits of armor? Dare we ask whether mothers, fathers, and children are all being shoe-horned into a family structure that suits none of us? Might the contemporary pandemics of fracturing families, parental exhaustion, and confused, resentful children be predictable consequences of what is, in truth, a distorted and distorting family structure inappropriate for our species?

Nuclear Meltdown

If the independent, isolated nuclear family unit is, in fact, the structure into which human beings most naturally configure themselves, why do contemporary societies and religions find it necessary to prop it up with tax breaks and supportive legislation while fiercely defending it from same-sex couples and others proposing to marry in supposedly "nontraditional" ways? One wonders, in fact, why marriage is a legal issue at all—apart from its relevance to immigration and property laws. Why would something so integral to human nature require such vigilant legal protection?

Furthermore, if the nuclear triad is so deeply embedded in our nature, why are fewer and fewer of us choosing to live that way? In the United States, the percentage of nuclear family households has dropped from 45 to 23.5 since the 1970s. Married couples (with and without children) accounted for roughly 84 percent of all American households in 1930, but the latest figure is just under 50 percent, while the number of unmarried couples living together has mushroomed from about 500,000 in 1970 to more than ten times that number in 2008.

Before Bronislaw Malinowski (1884–1942), the most respected and influential anthropologist of his day, declared the issue settled, there was plenty of debate over whether or not the mother-father-child triad was, in fact, the universal atomic unit of human social organization. Malinowski scoffed at Morgan's notion that societies could ever have been organized along nonnuclear lines, writing:

> These actors are *obviously* three in number at the beginning—the two parents and their offspring. . . . This unquestionably correct principle has become . . . the starting point for a new interpretation of Morgan's hypothesis of a primitive communal marriage. [They are] fully aware that group-marriage implies group-parenthood. Yet group-parenthood [is] an almost *unthinkable* hypothesis. . . . This conclusion has led to such capital howlers as that "the clan marries the clan and begets the clan" and that "the clan, like the family, is a reproductive group" [emphasis added].[12]

"Unquestioningly correct principle?" "Unthinkable hypothesis?" "Capital howlers?" Malinowski seems to have been personally offended that Morgan had dared to doubt the universality and naturalness of the sanctified nuclear family structure.

Meanwhile, within a few blocks of the London classrooms where he lectured, untold numbers of infants whose existence threatened to expose the colossal error at the heart of Malinowski's "unquestioningly correct principle" were being sacrificed, quite literally, in foundling hospitals. The situation was no less horrific in the United States. In 1915, a doctor named Henry Chapin visited ten foundling hospitals and found that in nine of them, every child died before the age of two. *Every* child.[13] This dark fate awaited inconvenient children born throughout Europe. In her memoir of middle-class life in early twentieth-century Germany, for instance, Doris Drucker describes the village "Angel-maker," who received babies from unwed mothers and "starved the little children in her care to death," while the unwed, now childless mother was hired out as a wet nurse to upper-class families.[14] How efficient.

Horrifying as it is to contemplate, widespread infanticide was not limited to Malinowski's day. For centuries, millions of European children had been passed through discreet revolving boxes set into the walls of foundling hospitals. These boxes were designed to protect the anonymity of the person leaving the child, but they offered scant protection to the infant. The survival rate in those institutions was little better than if the revolving boxes had opened directly into a crematorium's furnace. Far from being places of healing, these were government- and church-approved slaughterhouses where children whose existence might have raised inconvenient questions about the "naturalness" of the nuclear family were disposed of in a form of industrialized infanticide.[15]

In his book *Eve's Seed: Biology, the Sexes, and the Course of History*, historian Robert S. McElvaine gets off a few "capital howlers" of his own, writing, "the general trend in human evolution is *undeniably* toward pair bonding and lasting families. Pair bonding (albeit often with some backsliding, especially by men) and the family are," he insists, "the exceptions notwithstanding, among the traits that *characterize the human species* [emphasis added]."[16]

Sure, forget all the backsliding and the many exceptions, and you've got a real strong case!

Despite overwhelming evidence to the contrary, Malinowski's position remains deeply embedded in both scientific and popular assumptions about family structure. In fact, the whole architecture of what qualifies as *family* in Western society is based on Malinowski's insistence that each child everywhere has always had just one father.

But if Malinowski's position has won the day, why is poor Morgan's intellectual body still being regularly disinterred for further insult? Anthropologist Laura Betzig opens a paper on conjugal dissolution (failed marriage) by noting that Morgan's "fantasy [of group marriage] . . . expired on encountering the evidence, and a century after Morgan . . . the consensus is that [monogamous] marriage comes as close to being a human universal as anything about human behavior can."[17] Ouch. But in truth, Morgan's understanding of family structure was no "fantasy." His conclusions were based upon decades of extensive field research and study. Later, a bit less wind in her sails, Betzig admits that "there is still, however, no consensus as to why" marriage is so widespread.

That's a mystery all right. We'll see that anthropologists find marriage wherever they look mainly because they haven't quite decided what it looks like.

CHAPTER EIGHT

Making a Mess of Marriage, Mating, and Monogamy

Marriage is the most natural state of man, and therefore the state in which you are most likely to find solid Happiness.

BENJAMIN FRANKLIN

Love is an ideal thing, marriage a real thing; a confusion of the real with the ideal never goes unpunished.

JOHANN WOLFGANG VON GOETHE

When Albert Einstein proclaimed that $E=mc^2$, no physicists asked each other, "What's he mean by E?" In the hard sciences, the important stuff comes packaged in numbers and predefined symbols. Imprecise wording rarely causes confusion. But in more interpretive sciences such as anthropology, psychology, and evolutionary theory, misinterpretation and misunderstanding are common.

Take the words *love* and *lust*, for example. Love and lust are as different from each other as red wine and blue cheese, but because they can also complement one another splendidly, they get conflated with amazing, dumbfounding regularity.

In the literature of evolutionary psychology, in popular culture, in the tastefully appointed offices of marriage counselors, in religious teachings, in political discourse, and in our own mixed-up lives, lust is often mistaken for love. Perhaps even more insidious and damaging in societies insistent on long-term, sexually exclusive monogamy, the negative form of that statement is also true. The absence of lust is misread as indicating an absence of love (we'll explore this in Part V).

Experts inadvertently encourage us to confuse the two. Helen Fisher's *Anatomy of Love*, a book referenced earlier, is far more concerned with shared parental responsibility for a child's first few years than with the love joining the parents to one another. But we can't blame Fisher, as the language itself works against clarity. We can "sleep with" someone without ever closing our eyes.[1] When we read that the politician "made love" with the prostitute, we know love had little to do with it. When we report how many "lovers" we've had, are we claiming to have been "in love" with all of them? Similarly, if we "mate" with someone, does that make us "mates"? Show a guy a photo of a hot-looking woman and ask him if he'd like to "mate with her." Chances are good he'll say (or think), "Sure!" But chances are also high that marriage, children, and the prospect of a long future together never entered into his decision-making process.

Everyone knows these are arbitrary expressions for an almost infinite range of situations and relationships—everyone, it appears, but the experts. Many evolutionary psychologists and other researchers seem to think that "love" and "sex" are interchangeable terms. And they throw together "copulating" and "mating" as well. This failure to define terminology often leads to confusion and allows cultural bias to contaminate our thinking about human sexual nature. Let's try to hack a path through this tangled verbal undergrowth.

Marriage: The "Fundamental Condition" of the Human Species?

The intimate male-female relationship . . . which zoologists have dubbed a 'pair bond,' is bred into our bones. I believe this is what sets us apart from the apes more than anything else.

FRANS DE WAAL[2]

The majority of husbands remind me of an orangutan trying to play the violin.

HONORÉ DE BALZAC

The holy grail of evolutionary psychology is the "human universal." The whole point of the discipline is to tease out intrinsically *human* patterns of perception, cognition, and behavior from those determined on a cultural or personal level: Do you like baseball because you grew up watching games with Dad or because the sight of small groups of men strategizing and working together on a field connects to a primordial module in your brain? That's the sort of question evolutionary psychologists love to ask and aspire to answer.

Because evolutionary psychology is all about uncovering and elucidating the so-called *psychic unity of humankind*—and because of the considerable political and professional pressure to discover traits that conform to specific political agendas—readers need to be cautious about claims concerning such universals. Too often, the claims don't hold up to scrutiny.

The supposed universality of human marriage—and the linked omnipresence of the nuclear family—is a case in point. A cornerstone of the standard model of human sexual evolution, the claim for this universal human tendency to marry appears to be beyond question or doubt—"unquestioningly correct" in Malinowski's words. Though the tendency has been assumed since before Darwin, evolutionary biologist Robert Trivers's now-classic paper *Parental Investment and Sexual*

Selection, published in 1972, consolidated the position of marriage as foundational to most theories of human sexual evolution.[3]

Recall that marriage, as defined by these theories, represents the *fundamental exchange* underlying human sexual evolution. In his BBC television series *The Human Animal*, Desmond Morris flatly declares, "The pair bond is the fundamental condition of the human species." Michael Ghiglieri, biologist and protégé of Jane Goodall, writes, "Marriage . . . is the ultimate human contract. Men and women in all societies marry in nearly the same way. Marriage," continues Ghiglieri, "is normally a 'permanent' mating between a man and a woman . . . with the woman nurturing the infants, while the man supports and defends them. The institution of marriage," he concludes, "is older than states, churches, and laws."[4] Oh my. *The fundamental condition? The ultimate human contract?* Hard to argue with that.

But let's try, because slippery use of the word *marriage* in the anthropological literature has resulted in a huge headache for anyone trying to understand how marriage and the nuclear family *really* fit into human nature—if at all. The word, we'll find, is used to refer to a whole slew of different relationships.

In *Female Choices,* her survey of female primate sexuality, primatologist Meredith Small writes of the confusion that resulted when the term *consortship* drifted away from its original meaning—a striking parallel to the confusion over *marriage.* Small explains, "The word 'consortship' was used initially to define the close male-female sexual bond seen in savannah baboons and then usage of the word spread to the relationship of other mating pairs." This semantic leap, says Small, was a mistake. "Researchers began to think that all primates form consortships, and they applied the word to any short or long, exclusive or nonexclusive mating." This is a problem because "what was originally intended to describe a specific male-female association that lasted during the days surrounding ovulation became an all-inclusive word for mating. . . . Once a female is described as 'being in consort,' no one sees the importance of her regular copulations with other males."[5]

Biologist Joan Roughgarden has noted the same problematic application of present-day human mating ideals to animals. She writes, "Sexual selection's primary literature describes extrapair parentage as 'cheating' on the pair bond; the male is said to be 'cuckolded'; offspring of extrapair parentage are said to be 'illegitimate'; and females who do not participate in extrapair copulations are said to be 'faithful.' This judgmental terminology," concludes Roughgarden, "amounts to applying a contemporary definition of Western marriage to animals."[6]

Indeed, when familiar labels are applied, supporting evidence becomes far more visible than counter-evidence in a psychological process known as *confirmation bias*. Once we have a mental model, we're much more likely to notice and recall evidence supporting our model than evidence against it. Contemporary medical researchers attempt to neutralize this effect by using double-blind methodology in all serious research—where neither the researcher nor the subject knows which pills contain the real medicine.

Without a clear definition of what they're looking for, many anthropologists have found marriage wherever they've looked. George Murdock, a central figure in American anthropology, asserted in his classic cross-cultural anthropological survey that the nuclear family is a "universal human social grouping." He went on to declare that marriage is found in every human society.

But as we've seen, researchers trying to describe human nature are highly susceptible to Flintstonization: unconsciously tending to "discover" features that look familiar, and thereby universalizing contemporary social configurations while inadvertently blocking insight into the truth. Journalist Louis Menand noted this tendency in a piece in *The New Yorker*, writing, "The sciences of human nature tend to validate the practices and preferences of whatever regime happens to be sponsoring them. In totalitarian regimes, dissidence is treated as a mental illness. In apartheid regimes, interracial contact is treated as unnatural. In free-market regimes, self-interest is treated as hardwired."[7] Paradoxically, in each of these cases, so-called *natural* behavior has to be encouraged and *unnatural* aberrations punished.

The now-forgotten diseases *drapetomania* and *dysaethesia aethiopica* illustrate this point. Both were described in 1851 by Dr. Samuel Cartwright, a leading authority on the medical care of "Negroes" in Louisiana and a leading thinker in the pro-slavery movement. In his article "Diseases and Peculiarities of the Negro Race," Dr. Cartwright explained that *drapetomania* was the disease "causing Negroes to run away . . . the absconding from service" to their white owners, while *dysaethesia aethiopica* was characterized by "hebetude of and obtuse sensibility of the body." He noted that slave overseers often referred to this disease, more simply, as "rascality."[8]

Despite high-minded claims to the contrary—often couched in language chosen to intimidate would-be dissenters (*dysaethesia aethiopica*!)—science all too often grovels at the feet of the dominant cultural paradigm.

Another weakness of many of these studies is known as the "translation paradox:" the assumption that a word (*marriage*, for instance) translated from one language to another has an identical meaning.

We can agree that birds *sing* and bees *dance* only as long as we remember that their singing and dancing has almost nothing in common with ours—from motivation to execution. We use identical words to signify very different behaviors. It's the same with marriage.

People everywhere do pair off—even if just for a few hours, days, or years. Maybe they do it to share pleasure, to make babies, to please their families, to seal a political alliance or business deal, or just because they like each other. When they do, the resident anthropologist standing in the shadows of love says, "Aha, this culture practices marriage, too. It's universal!" But many of these relationships are as far from our sense of marriage as a string hammock is from Grandma's featherbed. Simply changing the jargon and referring to *long-term pair bond* rather than *marriage* is no better. As Donald Symons put it, "The lexicon of the English language is woefully inadequate to reflect accurately the texture of human experience. . . . To shrink the present vocabulary to one phrase—pair-bond—and to imagine that in so doing one is being scientific . . . is simply to delude oneself."[9]

On Matrimonial Whoredom

Even if we overlook the ubiquitous linguistic confusion, people who consider themselves to be married can have strikingly different notions of what their marriage involves. The Aché of Paraguay say that a man and woman sleeping in the same hut are married. But if one of them takes his or her hammock to another hut, they're not married anymore. That's it. The original no-fault divorce.

Among the !Kung San (also known as Ju/'hoansi) of Botswana, most girls *marry* several times before they settle into a long-term relationship. For the Curripaco of Brazil, marriage is a gradual, undefined process. One scientist who lived with them explains, "When a woman comes to hang her hammock next to her man and cook for him, then some younger Curripaco say they are married (*kainukana*). But older informants disagree; they say they are married only when they have demonstrated that they can support and sustain each other. Having a baby, and going through the fast together, cements a marriage."[10]

In contemporary Saudi Arabia and Egypt, there is a form of marriage known as *Nikah Misyar* (normally translated as "traveler's marriage"). According to a recent article from Reuters:

> Misyar appeals to men of reduced means, as well as men looking for a flexible arrangement—the husband can walk away from a misyar and can marry other women without informing his first wife. Wealthy Muslims sometimes contract misyar when on holiday to allow them to have sexual relations without breaching the tenets of their faith. Suhaila Zein al-Abideen, of the International Union of Muslim Scholars in Medina, said almost 80 percent of misyar

marriages end in divorce. "A woman loses all her rights. Even how often she sees her husband is decided by his moods," she said.[11]

In the Shia Muslim tradition, there is a similar institution called *Nikah Mut'ah* ("marriage for pleasure"), in which the relationship is entered into with a preordained termination point, like a car rental. These *marriages* can last anywhere from a few minutes to several years. A man can have any number of temporary wives at the same time (in addition to his "permanent wife"). Often used as a religious loophole in which prostitution or casual sex can fall within the bounds of religious requirements, there is no paperwork or ceremony required. Is this, too, *marriage*?

Apart from expectations of permanence or social recognition, what about virginity and sexual fidelity? Are they universal and integral parts of marriage, as parental investment theory would predict? No. For many societies, virginity is so unimportant there isn't even a word for the concept in their language. Among the Canela, explain Crocker and Crocker, "Virginity loss is only the first step into full marriage for a woman." There are several other steps needed before the Canela society considers a couple to by truly married, including the young woman's gaining social acceptance through her service in a "festival men's society." This premarital "service" includes sequential sex with fifteen to twenty men. If the bride-to-be does well, she'll earn payments of meat from the men, which will be paid directly to her future mother-in-law on a festival day.

Cacilda Jethá (coauthor of this book) conducted a World Health Organization study of sexual behavior among villagers in rural Mozambique in 1990. She found that the 140 men in her study group were involved with 87 women as wives, 252 other women as long-term sexual partners, and 226 additional women on an occasional basis, working out to an average of four ongoing sexual relationships per man, not counting the unreported casual encounters many of these men likely experienced as well.

Among the Warao, a group living in the forests of Brazil, ordinary relations are suspended periodically and replaced by ritual relations, known as *mamuse*. During these festivities, adults are free to have sex with whomever they like. These relationships are honorable and believed to have a positive effect upon any children that might result.

In his fascinating profile of the Pirahã and a scientist who studies them, journalist John Colapinto reports that "though [they] do not allow marriage outside their tribe, they have long kept their gene pool refreshed by permitting their women to sleep with outsiders."[12]

Among the Siriono, it's common for brothers to marry sisters, forming an altogether different sort of *Brady Bunch*. The marriage itself takes place without any sort of ceremony or ritual: no exchanges of property or vows, not even a feast. Just rehang your hammocks next to the women's and you're married, boys.

This casual approach to what anthropologists call "marriage" is anything but unusual. Early explorers, whalers, and fur trappers of the frigid north found the Inuit to be jaw-droppingly hospitable hosts. Imagine their confused gratitude when they realized the village headman was offering his own bed (wife included) to the weary, freezing traveler. In fact, the welcome Knud Rasmussen and others had stumbled into was a system of spouse exchange central to Inuit culture, with clear advantages in that unforgiving climate. Erotic exchange played an important role in linking families from distant villages in a durable web of certain aid in times of crisis. Though the harsh ecology of the Arctic dictated a much lower population density than the Amazon or even the Kalahari Desert, extra-pair sexual interaction helped cement bonds that offered the same insurance against unforeseen difficulties.

None of this behavior is considered adultery by the people involved. But then, *adultery* is as slippery a term as *marriage*. It's not just thy neighbor's wife who can lead a man astray, but thine own as well. A well-known moral guide of the Middle Ages, the *Speculum Doctrinale* (*Mirror of Doctrine*), written by Vincent of Beauvais, declared, "A man who loves his wife very much is an adulterer. Any love for someone else's wife, or too much love for one's own, is shameful." The author went on to advise, "The upright man should love his wife with his judgment, not his affections."[13] Vincent of Beauvais would have enjoyed the company of Daniel Defoe (of London), famous still as the author of *Robinson Crusoe.* Defoe scandalized Britain in 1727 with the publication of a nonfiction essay with the catchy title *Conjugal Lewdness: or, Matrimonial Whoredom.* Apparently that title was a bit much. For a later edition, he toned it down to *A Treatise Concerning the Use and Abuse of the Marriage Bed.* This was no desert island adventure but a moralizing lecture on the physical and spiritual dangers of enjoying sex with one's spouse.

Defoe would have appreciated the Nayar people, native to southern India, who have a type of marriage that doesn't necessarily include any sexual activity at all, has no expectation of permanence, and no cohabitation—indeed the bride may never see the groom again once the marriage ritual has been performed. But since divorce is not permitted in this system, the stability of these marriages must be exemplary, according to the anthropological surveys.

As these examples show, many qualities considered essential components of marriage in contemporary Western usage are anything but universal: sexual exclusivity, property exchange, even the intention to stay together for long. None of these are expected in many of the relationships evolutionary psychologists and anthropologists insist on calling marriage.

Now consider the confusion created by the words *mate* and *mating.* A mate sometimes refers to a sexual partner in a given copulation; other times, it refers to a partner in a recognized marriage, with whom children are raised and all sorts of behavioral and economic patterns are established. To mate with someone could mean to join together "till death do us part," or it could refer to nothing more than a quickie with "Julio down by the schoolyard." When evolutionary psychologists tell us that men and women have different innate cognitive or emotional "modules," which determine their reactions to a mate's infidelity, we suppose that this refers to a mate in a long-term relationship.

But you never know. When we read, "Sex differences in humans' mate-selection criteria exist and persist because the mechanisms that mediate mate evaluation differ for men and women," and that "a tendency to become sexually aroused by visual stimuli constitutes part of the mate-selection process in men,"[14] we scratch our heads, wondering whether this is a discussion of how people choose that special someone to introduce to Mom or merely the immediate, visceral response patterns heterosexual men often experience in the presence of an attractive woman. Given that men have shown these same response patterns in response to photographs, films, attractively attired mannequins, and a Noah's ark of farm animals—none of which are available for marriage—it seems that this language must refer to sexual attraction alone. But we're not really sure. At what point does a mate become a mate with whom to mate?

CHAPTER NINE

Paternity Certainty: The Crumbling Cornerstone of the Standard Narrative

According to anthropologist Robert Edgerton, the Marind-anim people of Melanesia believed:

> Semen was essential to human growth and development. They also married quite young, and to assure the bride's fertility, she had to be filled with semen. On her wedding night, therefore, as many as ten members of her husband's lineage had sexual intercourse with the bride, and if there were more men than this in the lineage, they had intercourse with her the following night. . . . A similar ritual was repeated at various intervals throughout a woman's life.[1]

Welcome to the family. Have you met my cousins?

Lest you think this a particularly unusual wedding celebration, it seems the ancestors of the Romans did something similar. Marriage was celebrated with a wedding orgy in which the husband's friends had intercourse with the bride, with witnesses standing by. Otto Kiefer, in his 1934 *Sexual Life in Ancient Rome*, explains that from the Roman perspective, "Natural and physical laws are alien and even opposed to the marriage tie. Accordingly, the woman who is entering marriage must atone to Mother Nature for violating her, and go through a period of free prostitution, in which she purchases the chastity of marriage by preliminary unchastity."[2]

In many societies, such unchaste shenanigans continue well beyond the wedding night. The Kulina of Amazonia have a ritual known as the *dutse'e bani towi*: the "order to get meat." Don Pollock explains that the village women "go in a group from household to household at dawn, singing to the adult men in each house, 'ordering' them to go hunting. At each house, one or more women in the group step forward to bang on the house with a stick; they will serve as the sex partners of the men of the house that night, if they are successful in their hunt. Women in the group . . . are not allowed to select their own husband."

What happens next is significant. Feigning reluctance, the men drag themselves from their hammocks and head off into the jungle, but before splitting up to hunt independently, they agree on a time and place outside the village to meet later, where they'll redistribute whatever they've bagged, thus ensuring that every man returns to the village with meat, guaranteeing extra-pair sex for one and all. Yet another nail in the coffin of the standard narrative.

Pollock's description of the hunters' triumphal return is beyond improvement:

> At the end of the day the men return in a group to the village, where the adult women form a large semicircle and sing erotically provocative songs to the men, asking for their 'meat.' The men drop their catch in a large pile in the middle of the semicircle, often hurling it down with dramatic gestures and smug smiles. . . . After cooking the meat and eating, each woman retires with the man whom she selected as her partner for the sexual tryst. Kulina engage in this ritual with great humor and perform it regularly.[3]

We'll bet they do. Pollock kindly confirmed our hunch that the Kulina word for "meat" (*bani*) refers both to food and to what you're thinking it does, dear reader. Maybe marriage isn't a human universal, but the capacity for sexual double entendre just might be.

Love, Lust, and Liberty at Lugu Lake

> *There is not now, and never has been, a society in which confidence in paternity is so low that men are typically more closely related genetically to their sisters' than to their wives' offspring. Happily promiscuous, nonpossessive, Rousseauian chimpanzees turned out not to exist; I am not convinced by the available evidence that such human beings exist either.*
>
> DONALD SYMONS, *The Evolution of Human Sexuality*

Symons's bold declaration was an expression of faith in parental investment theory and the central importance of paternity certainty in human evolution. But Symons was dead wrong on both points. As he wrote those ill-fated words in the late 1970s, primatologists in jungles along the Congo River were learning that bonobos are *precisely* the happily promiscuous and nonpossessive apes whose existence Symons declared impossible. And the Mosuo (pronounced MWO-swo—also referred to as *Na* or *Nari*), an ancient society in southwest China, are a society where paternity certainty is so low and inconsequential that men do indeed raise their sisters' children as their own.

> *Women and men should not marry, for love is like the seasons—it comes and goes.*
>
> YANG ERCHE NAMU (Mosuo woman)

In the mountains around Lugu Lake, near the border between China's Yunnan and Sichuan provinces, live about 56,000 people who enjoy a family system that has perplexed and fascinated travelers and scholars for centuries. The Mosuo revere Lugu Lake as the Mother Goddess, while the mountain towering over it, Ganmo, is respected as the Goddess of Love. Their language is rendered in Dongba, the sole pictographic language still used in the world today. They have no words for *murder*, *war*, or *rape*. The Mosuo's relaxed and respectful tranquility is accompanied by a nearly absolute sexual freedom and autonomy for both men and women.[4]

In 1265, Marco Polo passed through the Mosuo region and later recalled their unashamed sexuality, writing, "They do not consider it objectionable for a foreigner, or any other man, to have his way with their wives, daughters, sisters, or any other women in their home. They consider it a great benefit, in fact, saying that their gods and idols will be disposed in their favor and offer them material goods in great abundance. This is why they are so generous with their women toward foreigners." "Many times," wrote Polo, with a wink and a nudge, "a foreigner has wallowed in bed for three or four days with a poor sap's wife."[5]

Macho Italian that he was, Polo completely misread the situation. He misinterpreted the women's sexual availability as a commodity controlled by the men, when in fact, the most striking feature of the Mosuo system is the fiercely defended sexual autonomy of all adults, women as well as men.

The Mosuo refer to their arrangement as *sese*, meaning "walking." True to form, most anthropologists miss the point by referring to the Mosuo system as "walking marriage," and including the Mosuo on their all-encompassing lists of cultures that practice "marriage." The Mosuo themselves disagree with this depiction of their system. "By any stretch of the imagination, *sese* are not marriages," says Yang Erche Namu, a Mosuo woman who published a memoir about her childhood along the shores of Mother Lake. "All *sese* are of the visiting kind, and none involves the exchange of vows, property, the care of children, or expectations of fidelity." The Mosuo language has no word for *husband* or *wife*, preferring the word *azhu*, meaning "friend."[6]

The Mosuo are a matrilineal, agricultural people, passing property and family name from mother to daughter(s), so the household revolves around the women. When a girl reaches maturity at about thirteen or fourteen, she receives her own bedroom that opens both to the inner courtyard of the house and to the street through a private door. A Mosuo girl has complete autonomy as to who steps through this private door into her *babahuago* (flower room). The only strict rule is that her guest must be gone by sunrise. She can have a different lover the following night—or later that same night—if she chooses. There is no expectation of commitment, and any child she conceives is raised in her mother's house, with the help of the girl's brothers and the rest of the community.

Recalling her childhood, Yang Erche Namu echoes Malidoma Patrice Somé's description of his African childhood, explaining, "We children could roam at our own will and visit from house to house and village to village without our mothers' ever fearing for our safety. Every adult was responsible for every child, and every child in turn was respectful of every adult."[7]

Among the Mosuo, a man's sisters' children are considered his paternal responsibility—not those who may (or may not) be the fruit of his own nocturnal visits to various flower rooms. Here we see another society in which male parental investment is unrelated to biological paternity. In the Mosuo language, the word *Awu* translates to both *father* and *uncle.* "In place of one father, Mosuo children have many uncles who take care of them. In a way," writes Yang Erche Namu, "we also have many mothers, because we call our aunts by the name *azhe Ami*, which means 'little mother.'"[8]

In a twist that should send many mainstream theorists into a tailspin, sexual relations are kept strictly separate from Mosuo family relations. At night, Mosuo men are expected to sleep with their lovers. If not, they sleep in one of the outer buildings, never in the main house with their sisters. Custom prohibits any talk of love or romantic relationships in the family home. Complete discretion is expected from everyone. While both men and women are free to do as they will, they're expected to respect one another's privacy. There's no kissing and telling at Lugu Lake.

The mechanics of the *açia* relationships, as they are referred to by Mosuo, are characterized by a sacred regard for each individual's autonomy—whether man or woman.[9] Cai Hua, a Chinese anthropologist and author of *A Society without Fathers or Husbands*, explains, "Not only do men and women have the freedom to foster as many *açia* relationships as they want and to end them as they please, but each person can have simultaneous relationships with several *açia*, whether it be during one night or over a longer period." These relationships are discontinuous, lasting only as long as the two people are in each other's presence. "Each visitor's departure from the woman's home is taken to be the end of their *açia* relationship," according to Cai Hua. "There is no concept of *açia* that applies to the future. The *açia* relationship . . . only exists instantaneously and retrospectively," although a couple may repeat their visits as often as they wish.[10]

Particularly libidinous Mosuo women and men unashamedly report having had hundreds of relationships. Shame, from their perspective, would be the proper response to promises of or demands for fidelity. A vow of fidelity would be considered inappropriate—an attempt at negotiation or exchange. Openly expressed jealousy, for the Mosuo, is considered aggressive in its implied intrusion upon the sacred autonomy of another person, and is thus met with ridicule and shame.

Sadly, hostility toward this free expression of female sexual autonomy is not limited to narrow-minded anthropologists and thirteenth-century Italian explorers. Although the Mosuo have no history of trying to export their system or convincing anyone else of the superiority of their approach to love and sex, they have long suffered outside pressure to abandon their traditional beliefs, which outsiders seem to find threatening.

Once the Chinese established full control of the area in 1956, government officials began making annual visits to lecture the people on the dangers of sexual freedom and convince them to switch to "normal" marriage. In a bit of dubious publicity reminiscent of *Reefer Madness*, Chinese government officials showed up one year with a portable generator and a film showing "actors dressed as Mosuo . . . who were in the last stages of syphilis, who had gone mad and lost most of their faces." The audience response was not what the Chinese officials expected: their makeshift cinema was burned to the ground. But the officials didn't give up. Yang Erche Namu recalls "meetings night after night where they harangued and criticized and interrogated. . . . [The Chinese officials] ambushed men on their way to their lovers' houses, they dragged couples out of their beds and exposed people naked to their own relatives' eyes."

When even these heavy-handed tactics failed to convince the Mosuo to abandon their system, government officials insisted on bringing (if not demonstrating) "decency" to the Mosuo. They cut off essential deliveries of seed grain and children's clothing. Finally, literally starved into submission, many Mosuo agreed to participate in government-sponsored marriage ceremonies, where each was given "a cup of tea, a cigarette, pieces of candy, and a paper certificate."[11]

But the arm-twisting had little lasting effect. Travel writer Cynthia Barnes visited Lugu Lake in 2006 and found the Mosuo system still intact, though under pressure from Chinese tourists who, like Marco Polo 750 years earlier, mistake the sexual autonomy of Mosuo women for licentiousness. "Although their lack of coyness draws the world's attention to the Mosuo," Barnes writes, "sex is not the center of their universe." She continues:

> I think of my parents' bitter divorce, of childhood friends uprooted and destroyed because Mommy or Daddy decided to sleep with someone else. Lugu Lake, I think, is not so much a kingdom of women as a kingdom of family—albeit one blessedly free of politicians and preachers extolling "family values." There's no such thing as a "broken home," no sociologists wringing their hands over "single mothers," no economic devastation or shame and stigma when parents part. Sassy and confident, [a Mosuo girl will] grow

up cherished in a circle of male and female relatives. . . . When she joins the dances and invites a boy into her flower room, it will be for love, or lust, or whatever people call it when they are operating on hormones and heavy breathing. She will not need that boy—or any other—to have a home, to make a "family." She already knows that she will always have both.[12]

The Mosuo approach to love and sex may well finally be destroyed by the hordes of Han Chinese tourists who threaten to turn Lugu Lake into a theme-park version of Mosuo culture. But the Mosuo's persistence in the face of decades—if not centuries—of extreme pressure to conform to what many scientists still insist is human nature stands as a proud, undeniable counter-example to the standard narrative.

Mosuo woman (Photo: Jim Goodman)

Mosuo women
(Photo: Sachi Cunningham/www.germancamera.com)

On the Inevitability of Patriarchy

Despite societies like the Mosuo's in which women are autonomous and play crucial roles in maintaining social and economic stability, and plentiful evidence from dozens of foraging societies in which females enjoy high status and respect, many scientists rigidly insist that all societies are and always have been patriarchal. In *Why Men Rule* (originally titled *The Inevitability of Patriarchy*), sociologist Steven Goldberg provides an example of this absolutist view, writing, "Patriarchy is universal. . . . Indeed, of all social institutions there is probably none whose universality is so totally agreed upon. . . . There is not, nor has there ever been, any society that even remotely failed to associate authority and leadership in suprafamilial areas with the male. There are no borderline cases."[13] Strong words. Yet, in 247 pages, Goldberg fails to mention the Mosuo even once.

Goldberg does mention the Minangkabau of West Sumatra, Indonesia, but only in an appendix, where he cites two passages from others' research. The first, dating to 1934, says that men are generally served food before women. From this, Goldberg concludes that males wield superior power in Minangkabau society. This is as logically consistent as concluding that Western societies must be matriarchal because men often hold doors open for women, allowing them to pass first. The second passage Goldberg cites is from a paper co-authored by anthropologist Peggy Reeves Sanday, suggesting that the Minangkabau men have some degree of authority in the application of various aspects of traditional law.

There are two big problems with Goldberg's application of Sanday's work. First, there is no inherent contradiction between claiming that a society is *not* patriarchal and yet that men *do* enjoy various types of authority. This is simply illogical: Van Gogh's famous painting *The Starry Night* is not a "yellow painting," though there is plenty of yel-

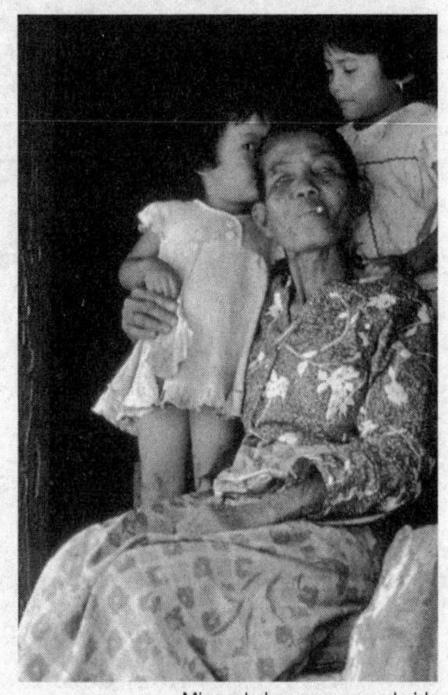

Minangkabau woman and girls (Photo: Christopher Ryan)[14]

low in it. The second problem with this citation is that Peggy Reeves Sanday, the anthropologist Goldberg cites, *has consistently argued that the Minangkabau are matriarchal.* In fact, her most recent book about the Minangkabau is called *Women at the Center: Life in a Modern Matriarchy.*[15]

Having spent over twenty summers living among the Minangkabau, Sanday says, "The power of Minangkabau women extends to the economic and social realms," noting, for example, that women control land inheritance and that a husband typically moves into the wife's household. The four million Minangkabau living in West Sumatra consider themselves to be a matriarchal society. "While we in the West glorify male dominance and competition," Sanday says, "the Minangkabau glorify their mythical Queen Mother and cooperation." She reports that "males and females relate more like partners for the common good than like competitors ruled by egocentric self-interest," and that as with bonobo social groups, women's prestige increases with age and "accrues to those who promote good relations. . . ."[16]

As happens so often in trying to understand and discuss other cultures, wording trips up specialists. When they claim never to have found a "true matriarchy," these anthropologists are envisioning a mirror image of patriarchy, a vision that ignores the differing ways males and females conceptualize and wield power. Sanday says that among the Minangkabau, for example, "Neither male nor female rule is possible because of [their] belief that decision-making should be by consensus." When she kept asking people which sex ruled, she was finally told that she was asking the wrong question. "Neither sex rules . . . because males and females complement one another."[17]

Remember this when some loudmouth at the bar declares that "patriarchy is universal, and always has been!" It's not, and it hasn't. But rather than feel threatened, we'd recommend that our male readers ponder this: Societies in which women have lots of autonomy and authority tend to be decidedly male-friendly, relaxed, tolerant, and plenty sexy. Got that, fellas? If you're unhappy at the amount of sexual opportunity in your life, don't blame the women. Instead, make sure they have equal access to power, wealth, and status. Then watch what happens.

As with bonobos, where female coalitions are the ultimate social authority and individual females need not fear the larger males, human societies in which women are "sassy and confident," as Barnes described the Mosuo girls—free to express their minds and sexuality without fear of shame or persecution—tend to be far more comfortable places for most men than societies ruled by a male elite. Maybe matriarchal societies are so difficult for Western male anthropologists to recognize because they expect a culture where men are suffering under the high heels of women—a reverse reflection of the long-standing male oppression of women in Western cultures. Instead, observing a society where most of the men are lounging about relaxed and happy, they conclude they've found yet another patriarchy, thereby missing the point entirely.

The March of the Monogamous

> *The idea of monogamy hasn't so much been tried and found wanting, as found difficult and left untried.*
>
> G. K. Chesterton

The surprise box-office hit of 2005 was a film called *March of the Penguins*. The second-biggest money-making documentary to date, viewers were touched by its depiction of the extreme dedication penguin couples showed in nurturing their adorable penguin chicks. Many viewers saw their own marriages reflected in the penguins' sacrifice for their offspring and for each other. As one reviewer put it, "It's impossible to watch the thousands of penguins huddled together against the icy Antarctic blasts . . . without feeling a tug of anthropomorphic kinship." Churches across the United States reserved cinemas for private screenings for their congregations. Rich Lowry, editor of *The National*

Review, told a conference of young Republicans, "Penguins are the really ideal example of monogamy. The dedication of these birds is amazing." Adam Leipzig, president of *National Geographic Feature Films*, declared the penguins "model parents," continuing, "What they go through to look after their children is phenomenal, and no parent who sees it will ever complain about the school run. There are parallels with human nature and it's moving to see."[18]

But unlike the birds themselves, penguin sexuality is not all black and white. That perfect penguin pair, that "ideal example of monogamy," those "model parents" are monogamous only as long as it takes to get their little one out of the egg, off the ice, and into the frigid Antarctic water—a little less than a year. If you've seen the film, you know that with all the trekking back and forth across the windswept ice and huddling against raging Antarctic blizzards, there's not much in the way of extramarital temptation anyway. Once Junior is swimming with the other eleven-month-olds—the penguin equivalent of kindergarten—fidelity is quickly forgotten, divorce is quick, automatic, and painless, and Mom and Dad are back on the penguin prowl. With a breeding adult typically living thirty years or more, these "model parents" have at least two dozen "families" in a lifetime. Did someone say "ideal example of monogamy"?

Whether you found the film to be cloyingly sweet or refreshingly so, a bold, if somewhat perverse, double feature would pair *The March of the Penguins* with Werner Herzog's *Encounters at the End of the World*. Herzog's documentary of the Antarctic is a masterpiece of photography and interviews with a range of surprising characters, including Dr. David Ainley, an almost comically reserved marine ecologist who has been studying penguins in Antarctica for two decades. Under Herzog's wry questioning, Ainley reports having witnessed cases of penguin ménages-à-trois, in which two males take turns caring for a particular female's egg, as well as "penguin prostitution," where females receive prime nest-building pebbles in exchange for a bit of penguin poontang.

The prairie vole is another supposed paragon of "natural monogamy." According to one newspaper article, "Prairie voles—squat rodents indigenous to plains and grasslands—are considered to be a near-perfect monogamous species. They form pair bonds that share a nest. Both male and female actively protect each other, their territory, and their young. The male is an active parent and, if one of the pair dies, the survivor does not take a new mate."[19] Considering the vitriol Darwin faced 150 years ago when he dared compare humans to apes, it's striking to note the scraps of comfort contemporary scientists find in equating human sexual behavior with that of the ratlike prairie vole. We who once compared ourselves to angels now see ourselves reflected in this lowly rodent. But C. Sue Carter and Lowell L. Getz, who have studied the biology of monogamy in prairie voles and other species for thirty-five years, are unambiguous: "Sexual exclusivity," they write "is not a feature of [the vole's] monogamy."[20] Thomas Insel, director of the National Institute of Mental Health (formerly director of Yerkes Primate Center) and an expert on the prairie vole, says that those in the know have a less exalted view of the prairie vole's monogamy: "They'll sleep with anyone but they'll only sit by their partners."[21]

Then there's that line (invariably directed at women, for some reason) that goes, "If you're looking for monogamy, marry a swan."*

So what about swans, then? Many species of birds have long been believed to be monogamous because two parents are needed for the 24/7 labor of incubating eggs and feeding nestlings. As with humans, investment-minded theorists assumed males would help out only if they were certain the young were their own. But the recent advent of affordable DNA testing has blown embarrassing holes in this story, too. Although a pair of bluebirds may build a nest and rear the young together, an average of 15 to 20 percent of the chicks are not sired by the male in the partnership, according to Patricia Adair Gowaty, a behavioral ecologist. And bluebirds aren't particularly slutty songbirds: DNA studies of the chicks of some 180 bird species previously thought

* Most famously in Nora Ephron's film *Heartburn*.

to be monogamous have shown that about 90 percent of them aren't. Swans, alas, are not among the virtuous 10 percent. So if you're looking for monogamy, forget the swan, too!

Is monogamy natural? Yes. . . . Human beings almost never have to be cajoled into pairing. Instead, we do this naturally. We flirt. We feel infatuation. We fall in love. We marry. And the vast majority of us marry only one person at a time. Pair-bonding is a trademark of the human animal.

HELEN FISHER

Strange trademark for a species that enjoys so much extra-pair sexual activity. The glue holding the standard narrative together is the assumption that *to marry* and *to mate* have universally applicable meanings, like the verbs *to eat*, or *to give birth*. But whatever terminology we use for the socially approved special relationship that often exists between men and women around the world will never communicate the universe of variations our species comes up with.

"Marriage," "mating," and "love" are socially constructed phenomena that have little or no transferable meaning outside any given culture. The examples we've noted of rampant ritualized group sex, mate-swapping, unrestrained casual affairs, and socially sanctioned sequential sex were all reported in cultures that anthropologists insist are monogamous simply because they've determined that something they call "marriage" takes place there. No wonder so many insist that marriage, monogamy, and the nuclear family are human universals. With such all-encompassing interpretations of the concepts, even the prairie vole, who "sleeps with anyone," would qualify.

CHAPTER TEN

Jealousy: A Beginner's Guide to Coveting Thy Neighbor's Spouse

[Once] marriage . . . becomes common, jealousy will lead to the inculcation of female virtue; and this, being honoured, will tend to spread to the unmarried females. How slowly it spreads to the male sex, we see at the present day.

CHARLES DARWIN[1]

In a traditional Canela marriage ceremony, the bride and groom lie down on a mat, arms under each other's heads, legs entwined. The brother of each partner's mother then comes forward. He admonishes the bride and her new husband to stay together until the last child is grown, specifically reminding them not be jealous of each other's lovers.

SARAH BLAFFER HRDY[2]

A printer's error in 1631 resulted in Bibles that proclaimed, "Thou shalt commit adultery."[3] Though not a biblical injunction, a common thread running through many of our examples of S.E.Ex. (Socio-Erotic Exchanges, if you've forgotten) is the explicit prohibition against relations with one's habitual partner(s), sometimes even under threat of death. Why would that be?

Since these rituals have developed in unrelated cultures throughout the world, they probably serve important functions. Internal conflict represented an existential threat to profoundly interdependent groups like those in which our ancestors lived for thousands of generations. Ritualized, socially sanctioned, sometimes even obligatory S.E.Ex. reduced disruptions caused by jealousy and possessiveness while blurring paternity. It's not surprising that small-scale societies highly dependent upon trust between individuals, generosity, and cooperation evolve and promote ways of enhancing these qualities while discouraging behavior and beliefs that would threaten group harmony and survival for group members.

It bears repeating that we are not attributing any particular nobility or, for that matter, ignobility to foragers. Some behaviors that seem normal to contemporary people (and which are therefore readily assumed to be universal) would quickly destroy many small-scale foraging societies, rendering them dysfunctional. Unrestrained self-interest, in particular, whether expressed as food-hoarding or excessive sexual possessiveness, is a direct threat to group cohesion and is therefore considered shameful and ridiculous.

Is there any doubt that societies can reshape such impulses?

Photo: Christopher White, www.christopherwhitephotography.com

Right now, girls' necks are being elongated ring by brass ring in parts of Thailand and Burma to make them more appealing to men. Clitorises are being cut away and labia sewn together in villages all over North Africa to dampen female desire, while in glamorous California, reduction labioplasty and other cosmetic vaginal surgeries have recently become a booming business. Elsewhere, the penises of boys are being circumcised or split open in ritualistic subincision. You get the point.

A few Native American tribes of the upper plains had an agreed-upon sense of beauty that led them to strap small planks of wood to their infants' still-pliable foreheads.[4] As the child grew, the straps would be tightened, as an orthodontist realigns a bite, bit by bit. It's unclear how much brain damage, if any, resulted from this practice, but the otherworldly conical heads that resulted scared the bejesus out of neighboring tribes and white fur trappers in the area.

Field sketch by Paul Kane[5]

And that may well have been the point, if you will. If their otherworldly appearance gave them a protective advantage by scaring potential enemies, it's not hard to see how such a fashion statement could have evolved. From savoring saliva beer or cow blood milkshakes to wearing socks with sandals, there is little doubt that people are willing to think, feel, wear, do, and believe pretty much anything if their society assures them it's *normal.*

Social forces that convince people to stretch their necks beyond the breaking point, schmush the heads of their infants, or sell their daughters into sacred prostitution are quite capable of reshaping or neutralizing sexual jealousy by rendering it silly and ridiculous. By rendering it *abnormal*.

The evolutionary explanation for male sexual jealousy, as we've seen, pivots on the genetic calculus underlying paternity certainty. But if it's a question of genes, a man should be far less concerned about his wife having sex with his brothers—who share half his genes—than with unrelated males. Gentlemen, would *you* be far less upset to find your wife in bed with your brother than with a total stranger? Ladies, would you prefer your husband have an affair with your sister? Didn't think so.[6]

Zero-Sum Sex

We mentioned David Buss in our discussion of mixed mating strategies earlier, but most of his work concerns the study of jealousy. Buss doesn't buy the notion of sharing food or mates, conceptualizing both in terms of scarcity: "If there is not enough food to feed all members of a group," he writes, "then some survive while others perish." Similarly, "If two women desire the same man . . . one woman's success in attracting him is the other woman's loss." Buss has little doubt that evolution is "a zero-sum game, with the victors winning at the expense of the losers."[7]

Far too often, the debate over the nature of human sexuality seems like a proxy war between antagonistic politico-economic philosophies. Defenders of the standard narrative see Cain's gain as Abel's loss, period. "That's just how life is, kid," they'll tell you. "It's human nature. Self-interest makes the world go round, pull yourself up by your bootstraps, it's a dog-eat-dog world and always has been."

This free-market vision of human mating hinges on the assumption that sexual monogamy is intrinsic to human nature. Absent monogamy (individual male "ownership" of female reproductive capacity), the I-win-you-lose dynamic collapses. As we outlined above, Buss and his colleagues get around the many glaring flaws in the theory (our extravagant sexual capacity, ubiquitous adultery in *all* cultures, rampant promiscuity in *both* our closest primate relatives, the absence of *any* monogamous primate living in large social groups) with pretzel logic and special pleading about *Homo sapiens*' internally conflicted, self-defeating "mixed mating strategies." Twist and stretch.

Buss and his colleagues have conducted scores of cross-cultural studies designed to confirm that men and women experience jealousy differently from each other, in consistent gender-specific ways. These researchers claim to have confirmed two important assumptions underlying the standard narrative: that men are universally worried about paternity certainty (hence, his mate's *sexual* fidelity is his main concern), while women are universally concerned with access to men's resources (so a woman will feel more threatened by any *emotional intimacy* that might inspire him to leave her for another woman). These gender-specific manifestations of sexual jealousy would appear to strongly support the standard narrative.

In a study typical of this research, Buss and his colleagues asked 1,122 people to imagine their partner becoming interested in someone else. They asked, "What would upset or distress you more: (a) imagining your partner forming a deep emotional (but not sexual) relationship with that person, or (b) imagining your partner enjoying a sexual (but not emotional) relationship with that person?" In studies like this conducted on college campuses around the United States and Europe, Buss and his colleagues consistently got more-or-less the same results. They found that men and women differed by roughly 35 percent in their responses, seeming to confirm their hypothesis. "Women continued to express greater upset about a partner's emotional infidelity," Buss writes, "even if it did not involve sex. Men continued to show more upset than women about a partner's sexual infidelity, even if it did not involve emotional involvement."[8]

But despite the apparent cross-cultural breadth of this research, it lacks methodological depth. Buss and his colleagues succumb to the same temptation that weakens so much sexuality research: reliance on a subject population more convenient than representative. Almost all the participants in these studies were university students. We understand that undergraduate students are low-hanging research fruit—easy for graduate students to locate and motivate (by offering partial course-credit for filling out a questionnaire, for example), but *this does not make them valid representatives of human sexuality.* Far from it. Even in supposedly liberal Western cultures, college-age people are in the early stages of their socio-sexual development with little, if any, experience to draw on when considering questions about one-night stands, long-term mate preferences, or their ideal number of lifetime sexual partners—all questions explored in Buss's research.

But Buss is not alone in this distorting focus on undergrads. The majority of research on sexuality is based upon the responses of eighteen- to twenty-two-year-old American university students. While one could make a case that a twenty-year-old guy is more or less like a turbo-charged fifty-year-old, few would argue that a twenty-year-old woman has much in common with a woman three decades older in terms of her sexuality. Most would agree that a woman's sexuality changes considerably throughout adulthood—for the better, if conditions allow.

Another problem with using college students in the sort of multicultural study Buss conducts concerns class distinctions. In underdeveloped countries, university students are likely to be from the upper classes. A wealthy Angolan student may have a lot more in common with a Portuguese undergrad than with someone his or her own age living in the slums of Luanda. Our own field research in Africa suggests that sexual beliefs and behavior differ greatly among social classes and subcultures there—as they do in other parts of the world.[9]

Beyond the distorting effects of age and class, Buss and his colleagues skip over the crucial fact that every one of their subjects lives in post-agricultural societies characterized by private property, political hierarchies, globalized television, and so on. How can we expect to identify "human universals" without including at least a few foragers, whose thoughts and behaviors have not been shaped by the effects of modern life and whose perspective represents the vast majority of our species' experience? As we've established, plenty of research on foragers demonstrates important similarities among unrelated societies and dramatic differences from post-agricultural norms. Swedes and upper-class Nigerians may see themselves as different from one another, but from a forager's perspective they would seem similar in many ways.

Granted, it is no easy matter to airdrop questionnaires and #2 pencils to foragers in the Upper Amazon (*The Grad Students Must Be Crazy*). Still, the difficulty or impossibility of including their perspective does nothing to lessen its vital importance to the integrity of this sort of research. This broad yet shallow research paradigm is like claiming to have uncovered "universal fish truths" after conducting studies in rivers around the world. What about the fish in lakes? Ponds? Oceans?

Psychologist Christine Harris has noted that Buss's conclusions could easily be nothing more than confirmation of old news: that "men are more reactive to any form of sexual stimuli than they are to emotional stimuli [and] are more interested in, or better able to imagine, such stimuli."[10] The men get more agitated by the sex, in other words, simply because they imagine it more clearly than the women do.

When Harris measured the bodily responses of people being asked Buss's questions, she found that "women as a group showed little difference in physiological reactivity," but they still predicted, almost unanimously, that the emotional infidelity would be more disturbing for them. This finding suggests a fascinating disconnect between what these women *actually* feel and what they think they *should* feel about their partner's fidelity (more on this later).

Psychologists David A. DeSteno and Peter Salovey found even more fundamental flaws in Buss's research, pointing out that the subjects' belief system comes into play when answering questions about hypothetical infidelity. They note that "the belief that emotional infidelity implies sexual infidelity was held to a greater degree by women than men," and that therefore, "the choice between sexual infidelity and emotional infidelity [at the heart of Buss's studies] is a false dichotomy. . . ."[11]

David A. Lishner and his colleagues honed in on another weak point: the fact that subjects are given only two options: either thoughts of sexual infidelity hurt more or thoughts of emotional infidelity do. Lishner asked, what if both scenarios made subjects feel *equally* uncomfortable? When Lishner included this third option, he found that the majority of respondents indicated that both forms of infidelity were equally upsetting, throwing further doubt on Buss's conclusions.[12]

Buss and other evolutionary psychologists who argue that some degree of jealousy is part of human nature may have a point, but they're overplaying their hand when they universalize their findings to everyone, everywhere, always. Human nature is made of highly reflective material. It is a mirror—admittedly marked by unalterable genetic scratches and cracks—but a mirror nonetheless. For most human beings, reality is pretty much what we're told it is. Like practically everything else, jealousy reflects social modification and can clearly be reduced to little more than a minor irritant if consensus deems it so.*

Among the Siriono of Bolivia, jealousy tends to arise not because one's spouse has lovers, but because he or she is devoting *too much time and energy* to those lovers. According to anthropologist Allan Holmberg, "Romantic love is a concept foreign to the Siriono. Sex, like hunger, is a drive to be satisfied." The expression *secubi* ("I like") is used in reference to everything the Siriono enjoy, whether food, jewelry, or a sexual partner. While "there are, of course, certain ideals of erotic bliss," Holmberg found that "under conditions of desire these readily break down, and the Siriono are content to conform to the principle of 'any port in a storm.'"[13]

* Real science offers one of the few—if not the only—reliable means of seeing beyond such cultural distortions, which makes it vitally important that we be fearless in rooting out cultural bias in research.

Anthropologist William Crocker is convinced that Canela husbands are not jealous, writing, "Whether or not Canela husbands are telling the truth about not minding, they join with other members in encouraging their wives to honor the custom . . . [of] ritual sex with twenty or more men during all-community ceremonies." Now, anyone who can pretend not to be jealous as his wife has sex with twenty or more men is someone you do *not* want to meet across a poker table.

The cultures we've reviewed, from steamy jungles in Brazil to lakeside Himalayan foothills, have each developed mechanisms for minimizing jealousy and sexual possessiveness. But the opposite also happens. Some cultures actively *encourage* the impulse toward possessiveness.

How to Tell When a Man Loves a Woman

Written by Percy Sledge and first recorded in 1966, "When a Man Loves a Woman" hit a cultural nerve. The song shot to the top of both the Billboard Hot 100 and R&B charts. Another version, recorded twenty-five years later by Michael Bolton, also went straight to the top of the charts, and the song now sits at number 54 on *Rolling Stone*'s list of the five hundred greatest songs of all time. Nothing is more prominent than love and sex in Western media, and "When a Man Loves a Woman" is an example of the message whispered in romantic ears throughout the world.

What does Mr. Sledge have to say about a man's love for a woman? What are the signs of true masculine love? Copyright restrictions won't allow us to quote the song's lyric in full, but most readers know the words by heart anyway. To review, when a man loves a woman:

- He becomes obsessed and can't think of anything else.
- He'll exchange anything, even the world, for her company.
- He's blind to any fault she may have, and will abandon even his closest friend if that friend tries to warn him about her.

- He'll spend all his money trying to hold her attention.
- And last but not least, he'll sleep in the rain if she tells him to.

We'd like to suggest an alternative title for this song: "When a Man Becomes Pathologically Obsessed and Sacrifices All Self-Respect and Dignity by Making a Complete Ass of Himself (and Losing the Woman Anyway Because Really, Who Wants a Boyfriend Who Sleeps Out in the Rain Because Someone Told Him To?)."

Similarly, "Every Breath You Take" sits at a respectable number 84 on *Rolling Stone*'s list of all-time great songs. One of the biggest hits of 1983, the song topped the U.K. charts for a month and the U.S. charts for two. It won Song of the Year, and The Police won that year's Grammy for Best Pop Performance. To date, the song has logged in over ten million registered air plays on radio stations around the world. Again, we're assuming you know the words. But have you ever *really* listened to them? Though often held up as one of the great love songs of all time, "Every Breath You Take" is not about love at all.

Sung from the perspective of a man who's been rejected by a woman who refuses to acknowledge that she "belongs" to him, he says he's going to follow her every step, watch her every move, see who she spends the night with, and so on.

This a *love song*? It should be #1 on Billboard's ranking of "Crazed & Dangerous Stalker Songs." Even Sting, who wrote the song after awakening in the middle of the night when the line "every breath you take / every move you make" bubbled up from his subconscious, didn't realize until later "how sinister [the song] is." He suggested in an interview that he may have been thinking of George Orwell's *1984*—a novel about surveillance and control—certainly *not* love.

So is jealousy *natural*? It depends. Fear is certainly natural, and like any other kind of insecurity, jealousy is an expression of fear. But whether or not someone else's sex life provokes fear depends on how sex is defined in a given society, relationship, and individual's personality.

First-born children often feel jealous when a younger sibling is born. Wise parents make a special point of reassuring the child that she'll always be special, that the baby doesn't represent any kind of threat to her status, and that there's plenty of love for everyone. Why is it so easy to believe that a mother's love isn't a zero-sum proposition, but that sexual love is a finite resource? Evolutionary biologist Richard Dawkins asks the pertinent question with characteristic elegance: "Is it so very obvious that you can't love more than one person? We seem to manage it with parental love (parents are reproached if they don't at least pretend to love all their children equally), love of books, of food, of wine (love of Château Margaux does not preclude love of a fine Hock, and we don't feel unfaithful to the red when we dally with the white), love of composers, poets, holiday beaches, friends . . . why is erotic love the one exception that everybody instantly acknowledges without even thinking about it?"[14]

Why, indeed? How would the prevalence and experience of jealousy be affected in Western societies if the economic dependence trapping most women and their children didn't exist, leading female sexual access to be a tightly controlled commodity? What if economic security and guilt-free sexual friendships were easily available to almost all men and women, as they are in many of the societies we've discussed, as well as among our closest primate cousins? What if no woman had to worry that a ruptured relationship would leave her and her children destitute and vulnerable? What if average guys knew they'd never have to worry about finding someone to love? What if we didn't all grow up hearing that *true love* is obsessive and possessive? What if, like the Mosuo, we revered the dignity and autonomy of those we loved? What if, in other words, sex, love, and economic security were as available to us as they were to our ancestors?

If fear is removed from jealousy, what's left?

Human beings will be happier—not when they cure cancer or get to Mars or eliminate racial prejudice or flush Lake Erie but when they find ways to inhabit primitive communities again. That's my utopia.

KURT VONNEGUT, JR.

According to E. O. Wilson, "all that we can surmise of humankind's genetic history argues for a more liberal sexual morality, in which sexual practices are to be regarded first as bonding devices and only second as a means for procreation."[15] We couldn't have said it better. But if human sexuality developed primarily as a bonding mechanism in interdependent bands where paternity certainty was a nonissue, then the standard narrative of human sexual evolution is toast. The anachronistic presumption that women have *always* bartered their sexual favors to individual men in return for help with child care, food, protection, and the rest of it collapses upon contact with the many societies where women feel no need to negotiate such deals. Rather than a plausible explanation for how we got to be the way we are, the standard narrative is exposed as contemporary moralistic bias packaged to look like science and then projected upon the distant screen of prehistory, rationalizing the present while obscuring the past. Yabba dabba doo.

PART III

The Way We Weren't

A central theme in our argument is that human sexual behavior is a reflection of both evolved tendencies and social context. Thus, a sense of the day-to-day social world in which human sexual tendencies evolved is essential to understanding them. It's hard to imagine the communal, cooperative social configurations we've described surviving long in the kind of world Hobbes envisioned, characterized by "*bellum omnium contra omnes*" (the war of all against all). But the false view of prehistoric human life summed up in Hobbes's pithy dictum "solitary, poor, nasty, brutish, and short" is still almost universally accepted.

Having established that prehistoric human life was highly social and decidedly *not* solitary, we now briefly address the other elements in Hobbes's description in the following four chapters, before continuing with a more direct discussion of overtly sexual material. We hope readers primarily interested in sex will bear with us because what might at first seem a detour is in fact a shortcut to a clearer vision of the day-to-day lives of our ancestors, a vision that will help you make better sense of the material that follows, as well as of your own world.

CHAPTER ELEVEN

"The Wealth of Nature" (Poor?)

The point is, ladies and gentleman, that greed, for lack of a better word, is good. Greed is right, greed works. Greed clarifies, cuts through, and captures the essence of the evolutionary spirit. Greed, in all of its forms . . . has marked the upward surge of mankind.

"Gordon Gekko," in the film *Wall Street*

What constitutes misuse of the universe? This question can be answered in one word: greed. . . . Greed constitutes the most grievous wrong.

Laurenti Magesa,
African Religion: The Moral Traditions of Abundant Life

Economics, "the dismal science," was dismal right from the start.

On a late autumn afternoon in 1838, what may have been the brightest bolt of illumination ever to flash out of an overcast English sky struck Charles Darwin right upside the head, leaving him stunned by what Richard Dawkins has called "the most powerful idea that has ever occurred to a man." At the very moment the great insight underlying natural selection came to him, Darwin was reading *An Essay on the Principle of Population* by Thomas Malthus.[1]

If the measure of an idea is its endurance through time, Thomas Malthus deserves his spot as Wikipedia's eightieth Most Influential Person in History. More than two centuries later, one would be hard pressed to find a single student of economics unfamiliar with the simple argument put forth by the world's first professor of economics. You'll recall that Malthus argued that each generation doubles geometrically (2, 4, 8, 16, 32 . . .), but farmers can only increase food supply arithmetically, as new fields are cleared and productive capacity is added in a linear fashion (2, 3, 4, 5, 6 . . .). From this crystalline reasoning follows Malthus's brutal conclusion: chronic overpopulation, desperation, and widespread starvation are intrinsic to human existence. Not a thing to be done about it. Helping the poor is like feeding London's pigeons; they'll just reproduce back to the brink of starvation anyway, so what's the point? "The poverty and misery which prevail among the lower classes of society," Malthus asserts, "are absolutely irremediable."

Malthus based his estimates of human reproductive rates on the recorded increase of (European) population in North America in the previous 150 years (1650–1800). He concluded that the colonial population had doubled every twenty-five years or so, which he took to be a reasonable estimate of the rates of human population growth in general.

In his autobiography, Darwin recalled that when he applied these dire Malthusian computations to the natural world, "it at once struck me that under these circumstances favorable variations would tend to be preserved, and unfavorable ones to be destroyed. The result of this would be the formation of new species. Here then I had at last got a theory by which to work . . ."[2] Science writer Matt Ridley believes Malthus taught Darwin the "bleak lesson" that "overbreeding must end in pestilence, famine or violence," convincing him that the secret of natural selection was embedded in the struggle for existence.

Thus was Darwin's brilliance sparked by the darkest Malthusian gloom.[3] Alfred Russel Wallace, who came up with the mechanism underlying natural selection independently of Darwin, experienced his own flash of insight while reading *the same essay* between bouts of fever in a hut on the banks of a malarial Malaysian river. Irish playwright George Bernard Shaw smelled the Malthusian morbidity underlying natural selection, lamenting, "When its whole significance dawns on you, your heart sinks into a heap of sand within you." Shaw lamented natural selection's "hideous fatalism," and complained of its "damnable reduction of beauty and intelligence, of strength and purpose, of honor and aspiration."[4]

But while Darwin and Wallace made excellent use of Malthus's dire calculations, there's a problem with them. They don't add up.

The tribes of hunters, like beasts of prey, whom they resemble in their mode of subsistence, will . . . be thinly scattered over the surface of the earth. Like beasts of prey, they must either drive away or fly from every rival, and be engaged in perpetual contests with each other. . . . The neighboring nations live in a perpetual state of hostility with each other. The very act of increasing in one tribe must be an act of aggression against its neighbors, as a larger range of territory will be necessary to support its increased numbers. . . . The life of the victor depends on the death of the enemy.

THOMAS MALTHUS, *An Essay on the Principle of Population*

If his estimates of population growth were even *close* to correct, Malthus (and thus, Darwin) would have been right to assume that human societies had long been "necessarily confined in room," resulting in "a perpetual state of hostility" with one another. In *Descent of Man*, Darwin revisits Malthus's calculations, writing, "Civilised populations have been known under favourable conditions, as in the United States, to double their numbers in twenty-five years . . . [At this] rate, the present population of the United States (thirty millions), would in 657 years cover the whole terraqueous globe so thickly, that four men would have to stand on each square yard of surface."[5]

If Malthus had been correct about prehistoric human population doubling every twenty-five years, these assumptions would indeed have been reasonable. But he wasn't, and they weren't. We now know that until the advent of agriculture, our ancestors' overall population doubled not every twenty-five years, but every *250,000 years*. Malthus (and thus, Darwin after him) was off by a factor of 10,000.[6]

Malthus assumed the suffering he saw around him reflected the eternal, inescapable condition of human and animal life. He didn't understand that the teeming, desperate streets of London circa 1800 were far from a reflection of prehistoric conditions. A century and a half earlier, Thomas Hobbes had made the same mistake, extrapolating from his own personal experience to conjure a mistaken vision of prehistoric human life.

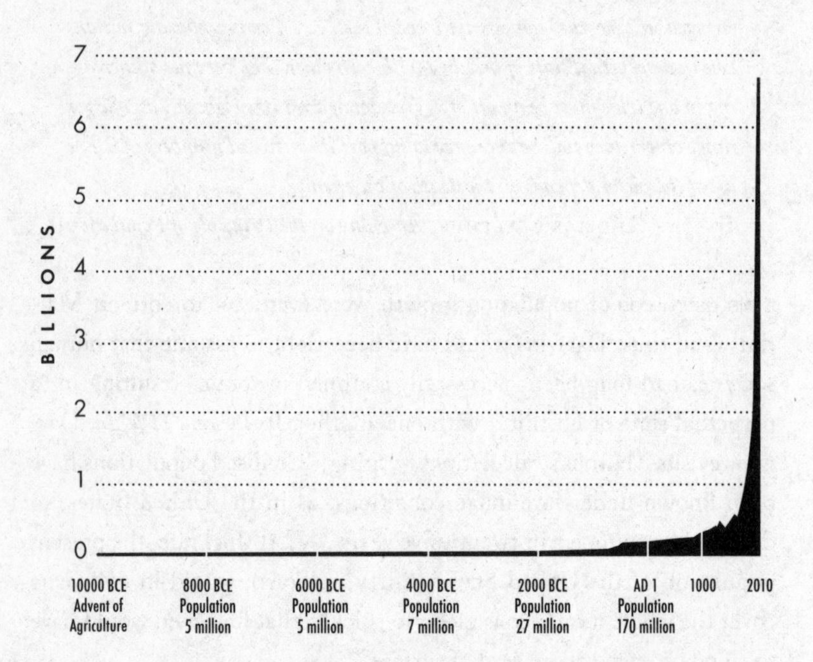

Thomas Hobbes was born to terror. His mother had gone into premature labor upon hearing that the Spanish Armada was about to attack England. "My mother," Hobbes wrote many years later, "gave birth to twins: myself and fear." *Leviathan*, the book in which he famously asserts that prehistoric life was "solitary, poor, nasty, brutish, and short," was composed in Paris, where he was hiding from enemies he'd made by supporting the Crown in the English Civil War. The book was nearly abandoned when he was taken with a near-fatal illness that left him at death's door for six months. Upon publication of *Leviathan* in France, Hobbes's life was now being threatened by his fellow exiles, who were offended by the anti-Catholicism expressed in the book. He fled back across the channel to England, begging the mercy of those he'd escaped eleven years earlier. Though he was permitted to stay, publication of his book was prohibited. The Church banned it. Oxford University banned it *and* burned it. Writing of Hobbes's world, cultural historian Mark Lilla describes "Christians addled by apocalyptic dreams [who] hunted and killed Christians with a maniacal fury they had once reserved for Muslims, Jews and heretics. It was madness."[8]

Hobbes took the madness of his age, considered it "normal," and projected it back into prehistoric epochs of which he knew next to nothing. What Hobbes called "human nature" was a projection of seventeenth-century Europe, where life for most was rough, to put it mildly. Though it has persisted for centuries, Hobbes's dark fantasy of prehistoric human life is as valid as grand conclusions about Siberian wolves based on observations of stray dogs in Tijuana.

To be fair, Malthus, Hobbes, and Darwin were constrained by the lack of actual data. To his enormous credit, Darwin recognized this and tried hard to address it—spending his entire adult life collecting specimens, taking copious notes, and corresponding with anyone who could provide him with useful information. But it wasn't enough. The necessary facts wouldn't be revealed for many decades.

But now we have them. Scientists have learned to read ancient bones and teeth, to carbon-date the ash of Pleistocene fires, to trace the drift of the mitochondrial DNA of our ancestors. And the information they've uncovered resoundingly refutes the vision of prehistory Hobbes and Malthus conjured and Darwin swallowed whole.

Poor, Pitiful Me

> *We are enriched not by what we possess, but by what we can do without.*
>
> Immanuel Kant

If George Orwell was correct that "those who control the past control the future," what of those who control the distant past?

Prior to the population increases associated with agriculture, most of the world was a vast, empty place in terms of human population. But the desperate overcrowding imagined by Hobbes, Malthus, and Darwin is still deeply embedded in evolutionary theory and repeated like a mantra, facts be damned. For example, in his recent essay entitled "Why War?," philosopher David Livingstone Smith projects the Malthusian panorama in all its mistaken despair: "Competition for limited resources is the engine of evolutionary change," he writes. "Any population that reproduces without inhibition will eventually outstrip the resources upon which it depends and, as numbers swell, individuals will have no alternative but to compete more and more desperately for dwindling resources. Those who can secure them will flourish, and those who cannot will die."[9]

True, as far as it goes. But it doesn't go very far, because Smith forgets that our ancestors were the original ramblin' men (and women)—

nomads who rarely stopped walking for more than a few days at a stretch. Walking away is what they did best. Why assume they would have stuck around to struggle "desperately" in an overpopulated area with depleted resources when they could simply walk up the beach, as they'd been doing for uncounted generations? And prehistoric human beings never reproduced "without inhibition," like rabbits. Far from it. In fact, prehistoric world population growth is estimated to have been well below .001 percent per year throughout prehistory[10]—hardly the population bomb Malthus assumed.

Basic human reproductive biology in a foraging context made rapid population growth unlikely, if not impossible. Women rarely conceive while breastfeeding, and without milk from domesticated animals, hunter-gatherer women typically breastfeed each child for five or six years. Furthermore, the demands of a mobile hunter-gatherer lifestyle make carrying more than one small child at a time unreasonable for a mother—even assuming lots of help from others. Finally, low body-fat levels result in much later menarche for hunter-gatherer females than for their post-agricultural sisters. Most foragers don't start ovulating until their late teens, resulting in a shorter reproductive life.[11]

Hobbes, Malthus, and Darwin were themselves surrounded by the desperate effects of population saturation (rampant infectious disease, ceaseless war, Machiavellian struggles for power). The prehistoric world, however, was sparsely populated—where it was populated at all. Other than isolated pockets surrounded by desert, or islands like Papua New Guinea, the prehistoric world was almost all open frontier. Most scholars believe that our ancestors were just setting out from Africa about fifty thousand years ago, entering Europe five or ten thousand years later.[12] The first human footprints probably weren't left on North American soil until about twelve thousand years ago.[13] During the many millennia before agriculture, the entire number of *Homo sapiens* on the planet probably never surpassed a million people and certainly never approached the current population of Chicago. Furthermore, recently obtained DNA analyses suggest several population bottlenecks caused by environmental catastrophes reduced our species to just a few thousand individuals as recently as 70,000 years ago.[14]

Ours is a very young species. Few of our ancestors faced the unrelenting scarcity-generated selective pressures envisioned by Hobbes, Malthus, and Darwin. The ancestral human journey did not, by and large, take place in a world already saturated with our kind, fighting over scraps. Rather, the route taken by the bulk of our ancestors led through a long series of ecosystems with nothing quite like us already there. Like the Burmese pythons recently set loose in the Everglades, cane toads spreading unchecked across Australia, or the timber wolves recently reintroduced to Yellowstone, our ancestors were generally entering an open ecological niche. When Hobbes wrote that "Man to Man is an arrant Wolfe," he was unaware of just how cooperative and communicative wolves can be if there's enough food for everyone. Individuals in species spreading into rich new ecosystems aren't locked in a struggle to the death against one another. Until the niche is saturated, such intraspecies conflict over food is counterproductive and needless.[15]

We've already shown that even in a largely empty world, the social lives of foragers were anything but solitary. But Hobbes also claimed prehistoric life was *poor,* and Malthus believed poverty to be eternal and inescapable. Yet most foragers don't believe themselves to be impoverished, and there's every indication that life wasn't generally much of a struggle for our fire-controlling, highly intelligent ancestors bound together in cooperative bands. To be sure, occasional catastrophes such as droughts, climatic shifts, and volcanic eruptions were devastating. But most of our ancestors lived in a largely unpopulated world, chock-full of food. For hundreds of thousands of generations, the omnivore's dilemma facing our ancestors lay in choosing among many culinary options. Plants eat soil; deer eat plants; cougars eat deer. But people can and do eat almost anything—including cougars, deer, plants, and yes, even soil.[16]

The Despair of Millionaires

Poverty . . . is the invention of civilization.

MARSHALL SAHLINS

A recent *New York Times* article under the headline "In Silicon Valley, Millionaires Who Don't Feel Rich" begins, "By almost any definition—except his own and perhaps those of his neighbors here in Silicon Valley—Hal Steger has it made." The article notes that although Mr. Steger and his wife have a net worth of roughly $3.5 million, he still typically works twelve-hour days plus another ten hours on weekends. "A few million," explains Steger, "doesn't go as far as it used to."

Gary Kremen (estimated net worth: $10 million), founder of Match.com, an online dating service, explains, "Everyone around here looks at the people above them." He continues to work sixty to eighty hours per week because, he says, "You're nobody here at $10 million." Another executive gets right to the point, saying, "Here, the top 1 percent chases the top one-tenth of 1 percent, and the top one-tenth of 1 percent chases the top one-one-hundredth of 1 percent."[17]

This sort of thinking isn't limited to Silicon Valley. A BBC report from September 2003 reported, "Well-off is the new poor." Dr. Clive Hamilton, a visiting scholar at Cambridge University, set out to study the "suffering rich" and found that four of every ten people earning over £50,000 (roughly $80,000 at the time) felt "deprived." Hamilton concluded, "The real concerns of yesterday's poor have become the imagined concerns of today's rich." Another recent survey in the United States found that 45 percent of those with a net worth (excluding their home) over $1 million were worried about running out of money before they died. Over one-third of those with more than $5 million had the same concern.[18]

"Affluenza" (a.k.a. luxury fever) is not an eternal affliction of the human animal, as some would have us believe. It is an effect of wealth disparities that arose with agriculture. Still, even in modern societies, we sometimes find echoes of the ancient egalitarianism of our ancestors.

In the early 1960s, a physician named Stewart Wolf heard about a town of Italian immigrants and their descendants in northeast Pennsylvania where heart disease was practically unknown. Wolf decided to take a closer look at the town, Roseto. He found that almost no one under age fifty-five showed symptoms of heart disease. Men over sixty-five suffered about half the number of heart problems expected of average Americans. The overall death rate in Roseto was about one-third below national averages.

After conducting research that carefully excluded factors such as exercise, diet, and regional variables like pollution levels, Wolf and sociologist John Bruhn concluded that the major factor keeping folks in Roseto healthier longer was *the nature of the community itself.* They noted that most households held three generations, that older folks commanded great respect, and that the community disdained any display of wealth, showing a "fear of ostentation derived from an ancient belief among Italian villagers relating to *maloccio* (the evil eye). Children," Wolf wrote, "were taught that any display of wealth or superiority over a neighbor would bring bad luck."

Noting that Roseto's egalitarian social bonds were already breaking down in the mid-1960s, Wolf and Bruhn predicted that within a generation, the town's mortality rates would start to shift upward. In follow-up studies they conducted 25 years later, they reported, "The most striking social change was a widespread rejection of a long standing taboo against ostentation," and that "sharing, once typical of Roseto, had given way to competition." Rates of both heart disease and stroke had doubled in a generation.[19]

Among foragers, where property is shared, poverty tends to be a nonissue. In his classic book *Stone Age Economics,* anthropologist Marshall Sahlins explains that "the world's most primitive people have few possessions, *but they are not poor.* Poverty is not a certain small amount of goods, nor is it just a relation between means and ends; above all it is a relation between people. Poverty is a social status. As such it is the invention of civilization."[20] Socrates made the same point 2,400 years ago: "He is richest who is content with least, for contentment is the wealth of nature."

But the wealth of civilization is material. After reading every word of the *Old Testament,* journalist David Plotz was struck by its mercantile tone. "The overarching theme of the Bible," he wrote, "particularly of Genesis, is real estate. God is . . . constantly making land deals (and then remaking them, on different terms). . . . It's not just land that the Bible is obsessed with, but also portable property: gold, silver, livestock."[21]

Malthus and Darwin were both struck by the characteristic egalitarianism of foragers, the former writing, "Among most of the American tribes . . . so great a degree of equality prevailed that all the members of each community would be nearly equal sharers in general hardships of savage life and in the pressure of occasional famines."[22] For his part, Darwin recognized the inherent conflict between the capital-based *civilization* he knew and what he saw as the natives' self-defeating generosity, writing, "Nomadic habits, whether over wide plains, or through the dense forests of the tropics, or along the shores of the sea, have in every case been highly detrimental. . . . The perfect equality of all the inhabitants," he wrote, "will for many years prevent their civilization."[23]

Finding Contentment "at the Bottom of the Scale of Human Beings"

Looking for an example of the world's most downtrodden, pathetic, desperately poor "savages," Malthus cited "the wretched inhabitants of Tierra del Fuego" who had been judged by European travelers to be "at the bottom of the scale of human beings." Just thirty years later Charles Darwin was in Tierra del Fuego, observing these same people. He agreed with Malthus concerning the Fuegians, writing in his journal, "I believe if the world was searched, no lower grade of man could be found."

As chance would have it, Captain Robert FitzRoy of the *Beagle*—the ship on which Darwin was sailing—had picked up three young Fuegians on an earlier voyage, and brought them back to England to introduce them to the glories of British life and a proper Christian education. Now, after they'd experienced firsthand the superiority of civilized living, FitzRoy was returning them to their own people to serve as missionaries. The plan was for them to show the Fuegians the folly of their "savage" ways and help them join the civilized world.

But just a year after Jemmy, York, and Fuegia had been returned to their people at Woollya cove, near the base of what is now called Mount Darwin, the *Beagle* and her crew returned to find the huts and gardens the British sailors had built for the three Fuegians deserted and overgrown. Eventually, Jemmy appeared and explained that he and the other Christianized Fuegians had reverted to their former way of living. Darwin, overcome with sadness, wrote in his journal that he'd never seen "so complete & grievous a change" and that "it was painful to behold him." They brought Jemmy aboard the ship and dressed him for dinner at the captain's table, much relieved to see that he at least remembered how to use a knife and fork properly.

Captain FitzRoy offered to bring him back to England, but Jemmy declined, saying he had "not the least wish to return to England" as he was "happy and contented" with "plenty fruits," "plenty fish," and "plenty birdies."

Remember the Yucatán. What looks like even extreme poverty—"the bottom of the scale of human beings"—may contain unrecognizable forms of wealth. Recall the "starving" Australian Aboriginal people, happily roasting low-fat rats and noshing on juicy grubs as revolted Englishmen looked on, certain they were witnessing the last demented spasms of starvation. When we start *detribalizing*—peeling away the cultural conditioning that distorts our vision—"wealth" and "poverty" may reveal themselves where we least expect to find them.[24]

CHAPTER TWELVE

The Selfish Meme (Nasty?)

Richard Dawkins, author of *The Selfish Gene*, coined the term *meme* to refer to a unit of information that can spread through a community via learning or imitation the way a favored gene is replicated through reproduction. Just as egalitarianism and resource- and risk-sharing memes were favored in the prehistoric environment, the selfishness meme has flourished in most of the post-agricultural world. Even so, no less an authority on economics than Adam Smith insisted that sympathy and compassion come to human beings as naturally as self-interest.[1]

The faulty assumption that scarcity-based economic thinking is somehow the de-facto human approach to questions of supply, demand, and distribution of wealth has misled much anthropological, philosophical, and economic thought over the past few centuries. As economist John Gowdy explains, '"Rational economic behavior' is peculiar to market capitalism and is an embedded set of beliefs, not an objective universal law of nature. The myth of economic man explains the organizing principle of contemporary capitalism, nothing more or less."[2]

Homo Economicus

We have a greed, with which we have agreed. . .

"Society," by Eddie Vedder

Many economists have forgotten (or never understood) that their central organizing principle, *Homo economicus* (a.k.a. *economic man*), is a myth rooted in assumptions about human nature, not a bedrock truth upon which to base a durable economic philosophy. When John Stuart Mill proposed what he admitted to be "an arbitrary definition of man, as a being who inevitably does that by which he may obtain the greatest amount of necessaries, conveniences, and luxuries, with the smallest quantity of labour and physical self-denial,"[3] it's doubtful he expected his "arbitrary definition" to delimit economic thought for centuries. Recall Rousseau's words: "If I had had to choose my place of birth, I would have chosen a state in which everyone knew everyone else, so that neither the obscure tactics of vice nor the modesty of virtue could have escaped public scrutiny and judgment." Those who proclaim that greed is simply part of human nature too often leave context unmentioned. Yes, greed is part of human nature. But so is shame. And so is generosity (and not just toward genetic relatives). When economists base their models on their fantasies of an "economic man" motivated only by self-interest, they forget community—the all-important web of meaning we spin around each other—the inescapable context within which anything truly *human* has taken place.

One of the most cited thought experiments in game theory and economics is called The Prisoner's Dilemma. It presents such an elegant and simple model of reciprocity, some scientists refer to it as "the *E. coli* of social psychology." Here's how it works: Imagine that two suspects are arrested, but the police don't have enough evidence for a conviction. After the prisoners are separated, each gets the same offer: If you testify against your partner and he remains silent, you'll go free and he'll get the full ten-year sentence. If he fesses up but you don't, you'll do the time while he walks free. If neither of you talks, you'll both get six months. If you both talk, you'll both do five years. Each prisoner must choose to snitch or remain silent. Each is told the other won't know about his decision. How will the prisoners respond?

In the classic form of the game, participants almost always betray one another, as each sees the benefit of quick betrayal: talk first, and walk away free. But take that theoretical conclusion to a prison anywhere in the world and ask what happens to "rats." Theory finally caught up to reality when scientists decided to let players gain experience with the game and see whether their behavior changed over time. As Robert Axelrod explains in *The Evolution of Cooperation*, players soon learned that they had a better chance if they kept quiet and assumed that their partner would do the same. If their partner talked, he acquired a bad reputation and was punished, in a "tit-for-tat" pattern. Over time, those players with the more altruistic approach flourished, while those who acted only in their individual short-term interest met serious problems—a shiv in the shower, maybe.

The classic interpretation of the experiment took another blow when psychologist Gregory S. Berns and his colleagues decided to monitor female players with an MRI machine. Berns et al. were expecting to find that subjects would react most strongly to being cheated—when one tried to cooperate and the other "snitched." But that's not what they found. "The results really surprised us," Berns told Natalie Angier, of *The New York Times*. The brain responded most energetically to acts of cooperation: "The brightest signals arose in cooperative alliances and in those neighborhoods of the brain already known to respond to desserts, pictures of pretty faces, money, cocaine and any number of licit and illicit delights."[4]

Analyzing the brain scans, Berns and his team found that when the women cooperated, two parts of the brain, both responsive to dopamine, were activated: the anteroventral striatum and the orbitofrontal cortex. Both regions are involved in impulse control, compulsive behavior, and reward processing. Though surprised by what his team found, Berns found comfort in it. "It's reassuring," he said. "In some ways, it says that we're wired to cooperate with each other."

The Tragedy of the Commons

First published in the prestigious journal *Science* in 1968, biologist Garrett Hardin's paper "The Tragedy of the Commons" is one of the most reprinted articles ever to appear in a scientific journal. The authors of a recent World Bank Discussion Paper called it "the dominant paradigm within which social scientists assess natural resource issues," while anthropologist G. N. Appell says the paper "has been embraced as a sacred text by scholars and professionals."[5]

Well into the 1800s, much of rural England was considered commons—property owned by the king but available to everyone—like the open range in the western United States before the advent of barbed-wire fencing. Using the English commons as his model, Hardin purported to show what happens when a resource is communally owned. He reasoned that in "a pasture open to all . . . each herdsman will try to keep as many cattle as possible." Though destructive to the pasture, the herdsman's selfishness makes good economic sense from his personal perspective. Hardin wrote, "The rational herdsman [will conclude] that the only sensible course for him to pursue is to add another animal to his herd." This is the only rational choice because all will share the cost of the degradation to the land from overgrazing, while the profit gained from additional animals will be his alone. Since each individual herdsman will come to the same conclusion, the common ground will inevitably be overgrazed. "Freedom in a commons," Hardin concluded, "brings ruin to us all."

Like Malthus's thoughts on population growth relative to agricultural capacity, Hardin's argument was a hit because (1) it features an A+B=C simplicity that appears to be inarguably correct; and (2) it is useful in justifying seemingly heartless decisions by entrenched powers. Malthus's essay, for example, was often cited by British business and political leaders to explain their inaction in the face of widespread poverty in Britain, including the famine of the 1840s in which several million Irish people starved to death (and millions more fled to the

United States). Hardin's articulation of the folly of communal ownership has provided cover repeatedly to those arguing for the privatization of government services and the conquest of native lands.

One other thing Hardin's elegant argument has in common with that of Malthus: it collapses on contact with reality.

As Canadian author Ian Angus explains, "Hardin simply ignored what actually happens in a real commons: self regulation by the communities involved." Hardin missed the fact that in small rural communities where population density is low enough that each of the herdsmen knows the others (the actual case in the historical English commons and in ancestral foraging societies), any individual who tries to game the system is quickly found out and punished. Nobel Prize–winning economist Elinor Ostrom's studies of commons management in small-scale communities led her to conclude that, "all communities have some form of monitoring to gird against cheating or using more than a fair share of the resource."[6]

Despite how it's been spun by economists and others arguing against local resource management, the real *tragedy of the commons* doesn't pose a threat to resources controlled by small groups of interdependent individuals. Forget the commons. We need to confront the tragedies of the open seas, skies, rivers, and forests. Fisheries around the world are collapsing because no one has the authority, power, and motivation to stop international fleets from strip-mining waters everybody (and thus, nobody) owns. Toxins from Chinese smokestacks burning illegally mined Russian coal lodge in Korean lungs, while American cars burning Venezuelan petroleum melt glaciers in Greenland.

What allows these chain-linked tragedies is the absence of local, personal shame. The false certainty that comes from applying Malthusian economics, the prisoner's dilemma, and the tragedy of the commons to pre-agricultural societies requires that we ignore the fine-grain contours of life in small-scale communities where nobody "could have escaped public scrutiny and judgment," in Rousseau's words. These tragedies become inevitable only when the group size exceeds our species' capacity for keeping track of one another, a point that's come to be known as *Dunbar's number.* In primate communities, size definitely matters.

Noticing the importance of grooming behavior in social primates, British anthropologist Robin Dunbar plotted overall group size against the neocortical development of the brain. Using this correlation, he predicted that humans start losing track of who's doing what to whom when group size hits about 150 individuals. In Dunbar's words, "The limit imposed by neocortical processing capacity is simply on the number of individuals with whom a stable inter-personal relationship can be maintained."[7] Other anthropologists had arrived at the same number by observing that when group sizes grew much beyond that, they tend to split into two smaller groups. Writing several years before Dunbar's paper was published in 1992, Marvin Harris noted, "With 50 people per band or 150 per village, everybody knew everybody else intimately, so that the bonding of reciprocal exchange could hold people together. People gave with the expectation of taking and took with the expectation of giving."[8] Recent authors, including Malcolm Gladwell in his best-selling *The Tipping Point*, have popularized the idea of 150 being a limit to organically functioning groups.

Having evolved in small, intimate bands where everybody knows our name, human beings aren't very good at dealing with the dubious freedoms conferred by anonymity. When communities grow beyond the point where every individual has at least a passing acquaintance with everyone else, our behavior changes, our choices shift, and our sense of the possible and of the acceptable grows ever more abstract.

The same argument can be made concerning the tragic misunderstanding of human nature that underlies communism: community ownership doesn't work in large-scale societies where people operate in anonymity. In *The Power of Scale*, anthropologist John Bodley wrote: "The size of human societies and cultures matters because larger societies will naturally have more concentrated social power. Larger societies will be less democratic than smaller societies, and they will have an unequal distribution of risks and rewards."[9] Right, because the bigger the society is, the less functional shame becomes. When the Berlin Wall came down, jubilant capitalists announced that the essential flaw of communism had been its failure to account for human nature. Well, yes and no. Marx's fatal error was his failure to appreciate the importance of context. Human nature functions one way in the context of intimate, interdependent societies, but set loose in anonymity, we become a different creature. Neither beast is more nor less *human*.

Dreams of Perpetual Progress

> *He is a barbarian, and thinks that the customs of his tribe and island are the laws of nature.*
>
> GEORGE BERNARD SHAW, Caesar in *Caesar and Cleopatra,* Act II

Were we really born in the best possible time and place? Or is ours a random moment in infinity—just another among uncountable moments, each with its compensating pleasures and disappointments? Perhaps you find it absurd to even entertain such a question, to assume there's any choice in the matter. But there is. We all have a psychological tendency to view our own experience as standard, to see our community as *The People,* to believe—perhaps subconsciously—that *we* are the chosen ones, God is on our side, and our team deserves to win. To see the present in the most flattering light, we paint the past in blood-red hues of suffering and terror. Hobbes has been scratching this persistent psychological itch for several centuries now.

It is a common mistake to assume that evolution is a process of improvement, that evolving organisms are progressing toward some final, perfected state. But they, and we, are not. An evolving society or organism simply adapts over the generations to changing conditions. While these modifications may be immediately beneficial, they are not really *improvements* because external conditions never stop shifting.

This error underlies the assumption that *here and now* is obviously better than *there and then*. Three and a half centuries later, scientists still quote Hobbes, telling us how lucky we are to live after the rise of the state, to have avoided the universal suffering of our barbaric past. It's deeply comforting to think we're the lucky ones, but let's ask the forbidden question: How lucky are we really?

Ancient Poverty or Assumed Affluence?

Prehistoric humans did not habitually store food, but this doesn't mean they lived in chronic hunger. Studies of prehistoric human bones and teeth show ancient human life was marked by episodic fasts and feasts, but prolonged periods of starvation were rare. How do we know our ancestors weren't living at the brink of starvation?

When children and adolescents don't get adequate nutrition for as little as a week, growth slows in the long bones in their arms and legs. When their nutritional intake recovers and the bones begin to grow again, the density of the new bone growth differs from the interruption. X-rays reveal these telltale lines in ancient bones, known as *Harris lines*.[10]

Periods of more prolonged malnutrition leave signs on the teeth known as hypoplasias—discolored bands and small pits in the enamel surface, which can still be seen many centuries later in fossilized remains. Archaeologists find fewer Harris lines and dental hypoplasias in the remains of prehistoric hunter-gatherer populations than they do in the skeletons of settled populations who lived in villages dependent on cultivation for their food supply. Being highly mobile, hunter-gatherers were unlikely to suffer from prolonged starvation since in most cases, they could simply move to areas where conditions were better.

Approximately eight hundred skeletons from the Dickson Mounds in the lower Illinois Valley have been analyzed. They reveal a clear picture of the health changes that accompanied the shift from foraging to corn farming around 1200 AD. Archaeologist George Armelagos and his colleagues reported that the farmers' remains show a 50 percent increase in chronic malnutrition, and three times the incidence of infectious diseases (indicated by bone lesions) compared with the foragers who preceded them. Furthermore, they found evidence of increased infant mortality, delayed skeletal growth in adults, and a fourfold increase in porotic hyperostosis, indicating iron-deficiency anemia in more than half the population.[11]

Many have noted the strangely cavalier approach to food among foragers, who have nothing in the freezer. French Jesuit missionary Paul Le Jeune, who spent some six months among the Montagnais in present-day Quebec, was exasperated by the natives' generosity. "If my host took two, three, or four Beavers," wrote Le Jeune, "whether it was day or night, they had a feast for all neighboring Savages. And if those people had captured something, they had one also at the same time; so that, on emerging from one feast, you went to another, and sometimes even to a third and a fourth." When Le Jeune tried to explain the advantages of saving some of their food, "They laughed at me. 'Tomorrow' (they said) 'we shall make another feast with what we shall capture.'"[12] Israeli anthropologist Nurit Bird-David explains, "Just as Westerners' behaviour is understandable in relation to their assumption of *shortage*, so hunter-gatherers' behaviour is understandable in relation to their *assumption of affluence*. Moreover, just as we analyze, even predict, Westerners' behavior by presuming that they behave as if they did not have enough, so we can analyze, even predict, hunter-gatherers' behaviour by presuming that they behave as if they had it made [emphasis added]."[13]

While farmers toil to grow rice, potatoes, wheat, or corn, a forager's diet is characterized by a variety of nutritious plants and critters. But how much work is foraging? Is it an efficient way to get a meal?

Archaeologist David Madsen investigated the energy efficiency of foraging for Mormon crickets (*Anabrus simplex*), which had been on the menu of the local native people in present-day Utah. His group collected crickets at a rate of about eighteen crunchy pounds per hour. At that rate, Madsen calculated that *in just an hour's work*, a forager could collect the caloric equivalent of eighty-seven chili dogs, forty-nine slices of pizza, or forty-three Big Macs—without all the heart-clogging fats and additives.[14] Before you scoff at the culinary appeal of Mormon crickets, give some thought to the frightening reality lurking within a typical chili dog. Another study found that the !Kung San (in the Kalahari desert, mind you) had an average daily intake (in a good month) of 2,140 calories and ninety-three grams of protein. Marvin Harris puts it simply: "Stone age populations lived healthier lives than did most of the people who came immediately after them."[15]

And maybe healthier than people who came *long* after them, too. The castles and museums of Europe are full of suits of armor too small to fit any but the most diminutive of modern men. While our medieval ancestors were shrimpy by modern standards, archaeologist Timothy Taylor believes that the human ancestors who first controlled fire—about 1.4 million years ago—were taller than the average person today. Skeletons dug up in Greece and Turkey show that pre-agricultural men in those areas were about five foot nine on average, with women being about five foot five. But with the adoption of agriculture, average height plummeted. Modern Greeks and Turks still aren't as tall, on average, as their ancient ancestors.

Throughout the world, the shift to agriculture accompanied a dramatic drop in the quality of most people's diets and overall health. Describing what he terms "the worst mistake in human history," Jared Diamond writes, "Hunter-gatherers practiced the most successful and longest-lasting life style in human history. In contrast," he concludes, "we're still struggling with the mess into which agriculture has tumbled us, and it's unclear whether we can solve it."

On Paleolithic Politics

Prehistoric life involved a lot of napping. In his provocative essay "The Original Affluent Society," Sahlins notes that among foraging people, "the food quest is so successful that half the time the people do not seem to know what to do with themselves."[16] Even Australian Aborigines living in apparently unforgiving and empty country had no trouble finding enough to eat (as well as sleeping about three hours per afternoon in addition to a full night's rest). Richard Lee's research with !Kung San bushmen of the Kalahari Desert in Botswana indicates that they spend only about fifteen hours per week getting food. "A woman gathers on one day enough food to feed her family for three days, and spends the rest of her time resting in camp, doing embroidery, visiting other camps, or entertaining visitors from other camps. For each day at home, kitchen routines, such as cooking, nut cracking, collecting firewood, and fetching water, occupy one to three hours of her time. This rhythm of steady work and steady leisure is maintained throughout the year."[17]

A day or two of light work followed by a day or two off. How's that sound?

Because food is found in the surrounding environment, no one can control another's access to life's necessities in hunter-gatherer society. Harris explains that in this context, "Egalitarianism is . . . firmly rooted in the openness of resources, the simplicity of the tools of production, the lack of non-transportable property, and the labile structure of the band."[18]

When you can't block people's access to food and shelter, and you can't stop them from leaving, how can you control them? The ubiquitous political egalitarianism of foraging people is rooted in this simple reality. Having no coercive power, leaders are simply those who are followed—individuals who have earned the respect of their companions. Such "leaders" do not—cannot—*demand* anyone's obedience. This insight is not breaking news. In his *Lectures on Jurisprudence*, which was published posthumously in 1896, Adam Smith wrote, "In a nation of hunters there is properly no government at all. . . . [They] have agreed among themselves to keep together for their mutual safety, but they have no authority one over another."

It's not surprising that conservative evolutionary psychologists have found foragers' insistence on sharing to be one of their most difficult nuts to crack. Given the iconic status of Dawkins's book *The Selfish Gene* and the popularized, status-quo protective notion of the all-against-all struggle for survival, the quest to explain why foraging people are so maddeningly generous to one another has occupied dozens of authors. In *The Origins of Virtue*, science writer Matt Ridley summarizes the inherent contradiction they face: "Our minds have been built by selfish genes, but they have been built to be social, trustworthy and cooperative."[19] One must walk a tightrope to insist that selfishness is (and has always been) the principal engine of human evolution even in the face of copious data demonstrating that human social organization was founded upon an impulse for sharing for many millennia.

Of course, this conflict would evaporate if proponents of the *always-selfish* theory of human nature accepted contextual limits to their argument. In other words, in a zero-sum context (like that of modern capitalist societies where we live among strangers), it makes sense, on some levels, for individuals to look out for themselves. But in other contexts human behavior is characterized by an equal instinct toward generosity and justice.[20]

Even if many of his followers prefer to ignore the subtleties of his arguments, Dawkins himself appreciates them fully, writing, "Much of animal nature is indeed altruistic, cooperative and even attended by benevolent subjective emotions. . . . Altruism at the level of the individual organism can be a means by which the underlying genes maximize their self-interest."[21] Despite famously inventing the concept of the "selfish gene," Dawkins sees group cooperation as a way to advance an individual's agenda (thereby advancing each individual's genetic interests). Why, then, are so many of his admirers unwilling to entertain the notion that cooperation among human beings and other animals may be every bit as *natural* and effective as short-sighted selfishness?

Nonhuman primates offer intriguing evidence of the "soft power of peace"—and not just horny bonobos, either. Frans de Waal and Denise Johanowicz devised an experiment to see what would happen when two different macaque species were placed together for five months. Rhesus monkeys (*Macaca mulatta*) are aggressive and violent, while stump-tails (*Macaca arctoides*) are known for their more chilled-out approach to life. The stump-tails, for example, make up after conflict by gripping each other's hips, whereas reconciliations are rarely witnessed among rhesus monkeys. Once the two species were placed together, however, the scientists saw that the more peaceful, conciliatory behavior of the stump-tails dominated the more aggressive rhesus attitudes. Gradually, the rhesus monkeys relaxed. As de Waal recounts, "Juveniles of the two species played together, groomed together, and slept in large, mixed huddles. Most importantly, the rhesus monkeys developed peacemaking skills on a par with those of their more tolerant group mates." Even when the experiment concluded, and the two species were once again housed only with their own kind, the rhesus monkeys were still three times more likely to reconcile after conflict and groom their rivals.[22]

A fluke? Neuroscientist/primatologist Robert Sapolsky has spent decades observing a group of baboons in Kenya, starting when he was a student in 1978. In the mid-1980s, a significant proportion of adult males in the group abruptly died of tuberculosis they'd picked up from infected food in a dump outside a tourist hotel. But the prized (albeit infected) dump food had been eaten only by the most belligerent baboons, who had driven away less aggressive males, females, or juveniles. Justice! With all the hard-ass males gone, the laid-back survivors were in charge. The defenseless troop was a treasure ready-made for pirates: a whole troop of females, sub-adults, and easily cowed males just waiting for some neighboring tough guys to waltz in and start raping and pillaging.

Because male baboons leave their natal troop at adolescence, within a decade of the dump cataclysm, none of the original, atypically mellow males were still around. But, as Sapolsky reports, "the troop's unique culture was being adopted by new males joining the troop." In 2004, Sapolsky reported that two decades after the tuberculosis "tragedy," the troop still showed higher-than-normal rates of males grooming and affiliating with females, an unusually relaxed dominance hierarchy, and physiological evidence of lower-than-normal anxiety levels among the normally stressed-out low-ranking males. Even more recently, Sapolsky told us that as of his most recent visit, in the summer of 2007, the troop's unique culture appeared to be intact.[23]

In *Hierarchy in the Forest*, primatologist Christopher Boehm argues that egalitarianism is an eminently rational, even hierarchical political system, writing, "Individuals who otherwise would be subordinated are clever enough to form a large and united political coalition, and they do so for the express purpose of keeping the strong from dominating the weak." According to Boehm, foragers are downright feline in refusing to follow orders, writing, "Nomadic foragers are universally—and all but obsessively—concerned with being free of the authority of others."[24]

Prehistory must have been a frustrating time for megalomaniacs. "An individual endowed with the passion for control," writes psychologist Erich Fromm, "would have been a social failure and without influence."[25]

What if—thanks to the combined effects of very low population density, a highly omnivorous digestive system, our uniquely elevated social intelligence, institutionalized sharing of food, casually promiscuous sexuality leading to generalized child care, and group defense—human prehistory was in fact a time of relative peace and prosperity? If not a "Golden Age," then at least a "Silver Age" ("Bronze Age" being taken)? Without falling into dreamy visions of paradise, can we—dare we—consider the possibility that our ancestors lived in a world where for most people, on most days, there was enough for everyone? By now, everyone knows "there's no free lunch." But what would it mean if our species evolved in a world where *every* lunch was free? How would our appreciation of prehistory (and consequently, of ourselves) change if we saw that our journey began in leisure and plenty, only veering into misery, scarcity, and ruthless competition a hundred centuries ago?

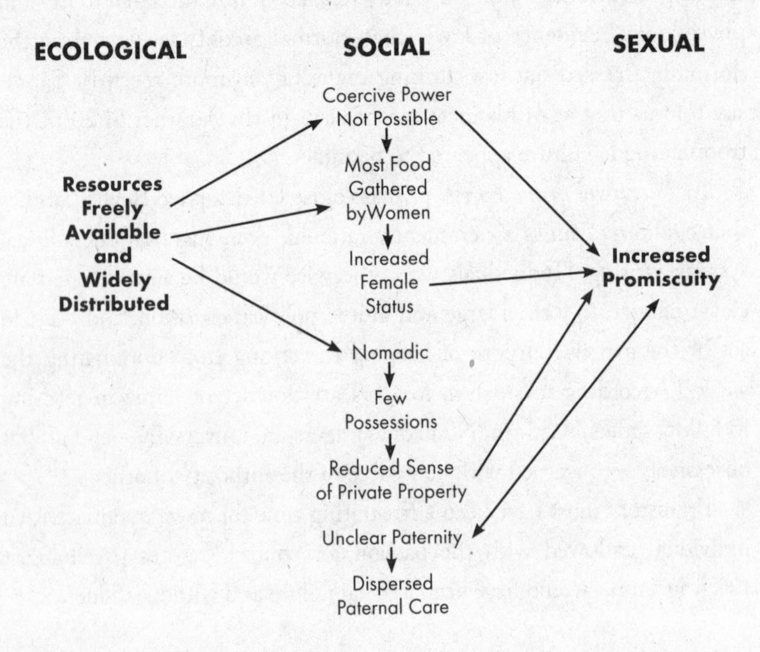

Difficult as it may be for some to accept, skeletal evidence clearly shows that our ancestors didn't experience widespread, chronic scarcity until the advent of agriculture. Chronic food shortages and scarcity-based economies are artifacts of social systems that arose with farming. In his introduction to *Limited Wants, Unlimited Means*, Gowdy points to the central irony: "Hunter-gatherers . . . spent their abundant leisure time eating, drinking, playing, socializing—in short, doing the very things we associate with affluence."

Despite no solid evidence to support it, the public hears little to dispute this apocalyptic vision of prehistory. The sense of human nature intrinsic to Western economic theory is mistaken. The notion that humans are driven only by self-interest is, in Gowdy's words, "a microscopically small minority view among the tens of thousands of cultures that have existed since *Homo sapiens* emerged some 200,000 years ago." For the vast majority of human generations that have ever lived, it would have been unthinkable to hoard food when those around you were hungry. "The hunter-gatherer," writes Gowdy, "represents *uneconomic man*."[26]

Remember, even those "wretched" inhabitants of Tierra del Fuego, condemned to "the bottom of the scale of human beings," threw down their hoes and walked away from their gardens once the HMS *Beagle* sailed out of sight. They knew firsthand how "civilized" people lived, yet they had "not the least wish to return to England." Why would they? They were "happy and contented" with "plenty fruits," "plenty fish," and "plenty birdies."

CHAPTER THIRTEEN

The Never-Ending Battle over Prehistoric War (Brutish?)

> *Evolutionists say that back in the twilight of life a beast, name and nature unknown, planted a murderous seed and that the impulse thus originated in that seed throbs forever in the blood of the brute's descendants. . . .*
>
> WILLIAM JENNINGS BRYAN[1]

Just as neo-Hobbesian fundamentalists hold that poverty is intrinsic to the eternal human condition, they maintain that war is fundamental to our nature. Author Nicolas Wade, for example, claims that "warfare between pre-state societies was incessant, merciless and conducted with the general purpose, often achieved, of annihilating the opponent."[2] According to this view, our propensity for organized conflict has roots reaching deep into our biological past, back to distant primate ancestors by way of our foraging forebears. It's *always* been about making war, supposedly, not love.

But nobody's very clear what all this incessant war was over. Despite his certainty that foragers' lives were plagued by "constant warfare," Wade acknowledges that "ancestral people lived in small egalitarian societies, without property, or leaders or differences of rank. . . ." So we're to understand that egalitarian, nonhierarchical, nomadic groups without property . . . were constantly at war? Hunter-gatherer societies, possessing so little and thus with so little to lose (other than their lives), living on a wide-open planet were nothing like the densely populated, settled societies struggling over dwindling or accumulated resources in more recent historical times.[3] Why would they be?

We've no space for a comprehensive response to this aspect of the standard Hobbesian narrative, but we've selected three well-known figures associated with it for a closer look at their arguments and data: evolutionary psychologist Stephen Pinker, the revered primatologist Jane Goodall, and the world's most famous living anthropologist, Napoleon Chagnon.[4]

Professor Pinker, Red in Tooth and Claw

Imagine a high-profile expert stands before a distinguished audience and argues that Asians are warlike people. In support of his argument, he presents statistics from seven countries: Argentina, Poland, Ireland, Nigeria, Canada, Italy, and Russia. "Wait a minute," you might say, "those aren't even Asian countries—except, possibly, Russia." The expert would be laughed off the stage—as he should be.

In 2007, world-famous Harvard professor and best-selling author Steven Pinker gave a presentation built upon similarly flawed logic at the TED conference (Technology, Entertainment, Design) in Long Beach, California.[5] Pinker's presentation provides both a concise statement of the neo-Hobbesian view of the origins of war and an illuminating look at the dubious rhetorical tactics often used to promote this bloodstained vision of our prehistory. The twenty-minute talk is available at the TED website.[6] We encourage you to watch at least the first five minutes (dealing with prehistory) before reading the following discussion. Go ahead. We'll wait here.

Though Pinker spends less than 10 percent of his time discussing hunter-gatherers (a social configuration, you'll recall, that represents well over 95 percent of our time on the planet), he manages to make a real mess of things.

Three and a half minutes into his talk, Pinker presents a chart based on Lawrence Keeley's *War Before Civilization: The Myth of the Peaceful Savage*. The chart shows "the percentage of male deaths due to warfare in a number of foraging or hunting and gathering societies." He explains that the chart shows that hunter-gatherer males were far more likely to die in war than are men living today.

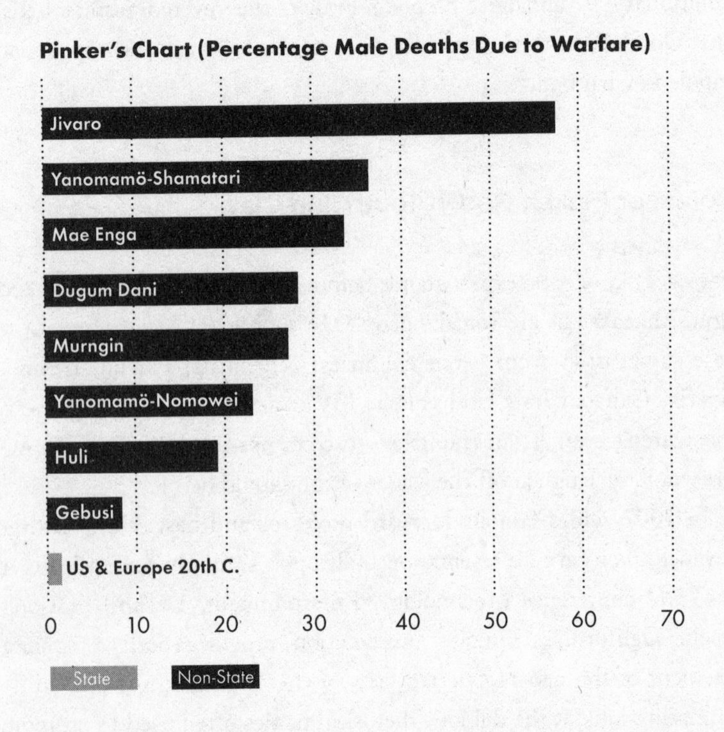

But hold on. Take a closer look at that chart. It lists seven "hunter-gatherer" cultures as representative of prehistoric war-related male death. The seven cultures listed are the Jivaro, two branches of Yanomami, the Mae Enga, Dugum Dani, Murngin, Huli, and Gebusi. The Jivaro and both Yanomami groups are from the Amazon region, the

Murngin are from northern coastal Australia, and the other four are all from the conflict-ridden, densely populated highlands of Papua New Guinea.

Are these groups representative of our hunter-gatherer ancestors? *Not even close.*[7]

Only *one* of the seven societies cited by Pinker (the Murngin) even approaches being an immediate-return foraging society (the way Russia is *sort of* Asian, if you ignore most of its population and history). The Murngin had been living with missionaries, guns, and aluminum powerboats for decades by the time the data Pinker cites were collected in 1975—not exactly prehistoric conditions.

None of the other societies cited by Pinker are immediate-return hunter-gatherers, like our ancestors were. They cultivate yams, bananas, or sugarcane in village gardens, while raising domesticated pigs, llamas, or chickens.[8] Even beyond the fact that these societies are not remotely representative of our nomadic, immediate-return hunter-gatherer ancestors, there are still further problems with the data Pinker cites. Among the Yanomami, true levels of warfare are subject to passionate debate among anthropologists, as we'll discuss shortly. The Murngin are not typical even of Australian native cultures, representing a bloody exception to the typical Australian Aborigine pattern of little to no intergroup conflict.[9] Nor does Pinker get the Gebusi right. Bruce Knauft, the anthropologist whose research Pinker cites on his chart, says the Gebusi's elevated death rates *had nothing to do with warfare.* In fact, Knauft reports that warfare is "rare" among the Gebusi, writing, "Disputes over territory or resources are extremely infrequent and tend to be easily resolved."[10]

Despite all this, Pinker stood before his audience and argued, with a straight face, that his chart depicted a fair estimate of typical hunter-gatherer mortality rates in prehistoric war. This is quite literally unbelievable.[11]

But Pinker is not alone in employing such sleight-of-hand to advance of Hobbes's dark view of human prehistory. In fact, this selective presentation of dubious data is disturbingly common in the literature on human blood-lust.

In their book *Demonic Males*, Richard Wrangham and Dale Peterson admit that war is unusual in nature, "a startling exception to the normal rule for animals." But because intergroup violence has been documented in both humans and chimps, they argue, a propensity for war must be an ancient human quality, going back to our last common ancestor. We are, they warn, "the dazed survivors of a continuous, 5-million-year habit of lethal aggression." Ouch.

But where are the bonobos? In a book of over 250 pages, the word *bonobo* appears on only eleven of them, and the species is dismissed as offering a less relevant sense of our last common ancestor than the common chimpanzee does—although many primatologists argue the opposite.[12] But at least they *mentioned* the bonobo.

In 2007, David Livingstone Smith, author of *The Most Dangerous Animal: Human Nature and the Origins of War*, published an essay exploring the evolutionary argument that war is rooted in our primate past. In his grisly accounts of chimps pummeling one another to a bloody pulp and eating each other alive, Smith repeatedly refers to them as "our closest non-human relative." You'd never know from reading his essay that we have an equally close nonhuman relative. The bonobo was left strangely—if typically—unmentioned.[13]

Amid the macho posturing about the brutal implications of chimpanzee violence, doesn't the *equally relevant*, nonwarring bonobo rate a mention, at least? Why all the yelling about yang with nary a whisper of yin? All darkness and no light may get audiences excited, but it can't illuminate them. This *oops-forgot-to-mention-the-bonobo* technique is distressingly common in the literature on the ancient origins of war.

But the bonobo's conspicuous absence is notable not just in discussions of war. Look for the missing bonobo wherever someone claims an ancient pedigree for human male violence of any sort. See if you can find the bonobo in this account of the origins of rape, from *The Dark*

Side of Man: "Men did not invent rape. Instead, they very likely inherited rape behavior from our ape ancestral lineage. Rape is a *standard* male reproductive strategy and likely has been one for millions of years. Male humans, chimpanzees, and orangutans *routinely* rape females. Wild gorillas violently abduct females to mate with them. Captive gorillas also rape females."[14] (Emphasis is in the original.)

Leaving aside the complications of defining *rape* in nonhuman species unable to communicate their experiences and motivations, rape—along with infanticide, war, and murder—has never been witnessed among bonobos in several decades of observation. Not in the wild. Not in the zoo. *Never.*

Doesn't that warrant a footnote, even?

The Mysterious Disappearance of Margaret Power

Even apart from doubts raised by bonobos, there are serious questions worth asking about the nature of chimp "warfare." In the 1970s, Richard Wrangham was a graduate student studying the relation between food supply and chimp behavior at Jane Goodall's research center at Gombe, Tanzania. In 1991, five years before Wrangham and Peterson's *Demonic Males* came out, Margaret Power published a carefully researched book, *The Egalitarians: Human and Chimpanzee*, that asked important questions concerning some of Goodall's research on chimpanzees (without, it must be said, ever expressing anything but admiration for Goodall's scientific integrity and intentions). But Power's name and her doubts are nowhere to be found in *Demonic Males.*

Power noticed that data Goodall collected in her first years at Gombe (from 1961 to 1965) painted a different picture of chimpanzee social interaction than the accounts of chimpanzee warfare she and her colleagues published to global acclaim a few years later. Observations from those first four years at Gombe had left Goodall with the impression that the chimps were "far more peaceable than humans." She saw no evidence of "war" between groups and only sporadic outbreaks of violence between individuals.

These initial impressions of overall primate peace mesh with research published four decades later, in 2002, by primatologists Robert Sussman and Paul Garber, who conducted a comprehensive review of the scientific literature on social behavior in primates. After reviewing more than eighty studies of how various primates spend their waking hours, they found that "in almost all species across the board, from diurnal lemurs—the most primitive primates—to apes . . . usually less than 5 percent of their day is spent in any active social behavior whatsoever." Sussman and Garber found that "usually less than 1 percent of their day is spent fighting or competing, and it's unusually much less than 1 percent." They found cooperative, affiliative behavior like playing and grooming to be ten to twenty times more common than conflict in all primate species.[15]

But Goodall's impression of relative harmony was to change—not coincidentally, argues Power—precisely when she and her students began giving the chimps hundreds of bananas every day, to entice them to hang around the camp so they could be observed more easily.

In the wild, chimps spread out to search for food individually or in small groups. Because the food is scattered throughout the jungle, competition is unusual. But, as Frans de Waal explains, "as soon as humans start providing food, even in the jungle, the peace is quickly disturbed."[16]

The mounds of deliciously smelly fruit locked in reinforced concrete boxes opened only for timed, regular feedings altered the chimps' behavior dramatically. Goodall's assistants had to keep rebuilding the boxes, as the frustrated apes found endless ways of prying or smashing them open. Ripe fruit that could not be eaten immediately was a new experience for them—one that left the chimps confused and enraged. Imagine telling a room of unruly three-year-olds on Christmas morning (each with the strength of four adult men) that they'll have to wait an unspecified amount of time to open the piles of presents they can see right there, under the tree.

Recalling this period a few years later, Goodall wrote, "The constant feeding was having a marked effect upon the behaviour of the chimps. They were beginning to move about in large groups more often than they had ever done in the old days. They were sleeping near camp and arriving in noisy hordes early in the morning. Worst of all, *the adult males were becoming increasingly aggressive*. . . . Not only was there *a great deal more fighting than ever before*, but many of the chimps were hanging around camp for hours and hours every day [emphasis added]."[17]

Margaret Power's doubts concerning Goodall's provisioning of the chimps have been largely left unaddressed by most primatologists, not just Wrangham.[18] Michael Ghiglieri, for example, went to study the chimps in Kibale Forest in nearby Uganda *specifically* in response to the notion that the intergroup conflict Goodall's team had witnessed might have been due to the distorting effects of those banana boxes. Ghiglieri writes, "My mission . . . [was] to find out whether these warlike killings were normal or an artifact of the researchers having provisioned the chimps with food to observe them."[19] But somehow Margaret Power's name doesn't even appear in the index of Ghiglieri's book, published eight years after hers.

We lack the space to adequately explore the questions Power raised, or to address subsequent reports of intergroup conflict among some (but not all) unprovisioned chimps in other study areas.[20] While we've got our doubts about the motivations of Pinker and Chagnon (see below), like Margaret Power, we have none about Jane Goodall's intentions or scientific integrity. Still, with all due respect to Goodall, Power's questions deserve consideration by anyone seriously interested in the debate over the possible primate origins of warfare.

The Spoils of War

Margaret Power's questions cut to the heart of the matter: why fight if there's nothing worth fighting over? Before the scientists started provisioning the apes, food appeared throughout the jungle, so the chimps spread out in search of something to eat each day. Chimps often call out to the others when they find a fruiting tree; mutual aid helps everyone, and feeding in the forest isn't a zero-sum endeavor. But once they learned that there would be a *limited* amount of easy food available in the same place each day, more and more chimps started arriving in aggressive, "noisy hordes" and "hanging around." Soon after, Goodall and her students began witnessing the now-famous "warfare" between chimp groups.

Perhaps for the first time ever, the chimps had something worth fighting over: a concentrated, reliable, yet limited source of food. Suddenly, they lived in a zero-sum world.

Applying this same reasoning to human societies, we're left wondering why immediate-return hunter-gatherers would risk their lives to fight wars. Over what, exactly? Food? That's spread out in the environment. Societies indigenous to areas where food is concentrated by natural conditions, like the periodic salmon runs of the Pacific Northwest of the United States and Canada, tend not to be immediate-return hunter-gatherers. We're more likely to find complex, hierarchical societies like the Kwakiutl (discussed later) in such spots. Possessions? Foragers have few possessions of any nonsentimental value. Land? Our ancestors evolved on a planet nearly empty of human beings for the vast majority of our existence as a species. Women? Possibly, but this claim presumes that population growth was important to foragers and that women were commodities to be fought over and traded like the livestock of pastoralists. It's likely that keeping population stable was more important to foragers than expanding it. As we've seen, when a group reaches a certain number of people, it tends to split into smaller groups anyway, and there is no inherent advantage in having *more* people to feed in band-level societies. We've also seen that women and men would have been free to move among different bands in the fission-fusion social system typical of hunter-gatherers, chimps, and bonobos.

The causal reverberations between social structure (foraging, horticultural, agrarian, industrial), population density, and the likelihood of war is supported by research conducted by sociologist Patrick Nolan, who found, "Warfare is more likely in advanced horticultural and agrarian societies than it is in hunting-and-gathering and simple horticultural societies." When he limited his analysis only to hunter-gatherer and agrarian societies, Nolan found that above-average population density was the best predictor of war.[21]

This finding is problematic for the argument that human war is a "5-million-year habit," given our ancestors' low population densities until the post-agricultural population explosion began just a few thousand years ago. Recent research looking at changes in mitochondrial DNA confirms that already low prehistoric global human population levels dropped nearly to extinction at several points (due to climatological catastrophes probably triggered by volcanic eruptions, asteroid strikes, and sudden changes in ocean currents). As mentioned previously, the entire world population of *Homo sapiens* may have dropped to just a few thousand individuals as recently as 74,000 years ago, when the massive Toba eruption severely disrupted world climate. But even with much of the northern hemisphere covered in ice, the world was anything but crowded for our distant ancestors.[22]

Population demographics have triggered wars in more recent historical times. Ecologist Peter Turchin and anthropologist Andrey Korotayev looked at data from English, Chinese, and Roman history, finding strong statistical correlations between increases in population density and warfare. Their research suggests population growth could account for as much as 90 percent of the variation between historical periods of war and peace.[23]

Early agriculture's stores of harvested grain and herds of placid livestock were like boxes of bananas in the jungle. There was now something worth fighting over: more. More land to cultivate. More women to increase population to work the land, raise armies to defend it, and help with the harvest. More slaves for the hard labor of planting, harvesting, and fighting. Failed crops in one area would lead desperate farmers to raid neighbors, who would retaliate, and so on, over and over.[24]

Freedom (from war) is just another word for nothing to lose—or gain.

But neo-Hobbesians ignore this rather straightforward analysis and the data supporting it, insisting that war *must* be an eternal human drive, all too often resorting to desperate rhetorical tactics like Pinker's to defend their view.

In the fourth chapter of his book *Sick Societies: Challenging the Myth of Primitive Harmony,* for example, Robert Edgerton writes, "Social stratification developed in some small-scale societies that lacked not only bureaucracies and priesthoods but cultivation as well." Okay, but in support of this assertion about social stratification and brutal rule by elites in "small-scale societies," he offers fifteen pages of vivid descriptions of, in this order (and leaving nothing out):

- the Kwakiutl Indians of Vancouver Island (a slave-owning, settled, property-accumulating, potlatch-celebrating, complex, hierarchical society);
- the Aztec Empire (numbering in the millions, with elaborate religious structures, priesthoods, and untold acres of slave-cultivated land around a capital city larger than any in Europe at the time of first contact, featuring sewage systems and lighted streets at night);
- the Zulu Empire (again, numbering well into the millions, with slavery, intensive agriculture, animal domestication, and continent-wide trade networks);
- the Asante Empire of present-day Ghana, which, Edgerton tells us, "was incomparably the greatest military power in West Africa."[25]

What any of these empires have to do with small-scale societies with no bureaucracies, priesthoods, or cultivation, Edgerton doesn't say. In fact, he doesn't mention a single foraging society for the rest of the chapter. This is like declaring that cats are difficult to train, then offering as evidence German shepherds, beagles, greyhounds, and golden retrievers.

In *Beyond War*, anthropologist Doug Fry rebuts the neo-Hobbesian view of universal war. "The belief that 'there has always been war,'" Fry writes, "does not correspond with the archaeological facts of the matter." Anthropologist Leslie Sponsel agrees, writing, "Lack of archaeological evidence for warfare suggests that it was rare or absent for most of human prehistory." After conducting a comprehensive review of prehistoric skeletal evidence, anthropologist Brian Ferguson concluded that apart from one particular site in modern-day Sudan, "only about a dozen *Homo sapiens* skeletons 10,000 years old or older, out of hundreds of similar antiquity examined to date, show clear indications of interpersonal violence." Ferguson continues, "If warfare were prevalent in early prehistoric times, the abundant materials in the archaeological record would be rich with evidence of warfare. But the signs are not there."[26]

Our bullshit detectors go off when scholars point to violent chimps and a few cherry-picked horticultural human societies mislabeled as foragers, claiming these as evidence of ancient tendencies toward warfare. Even more troubling, these scholars often remain mute on the distorting effects on chimps of food provisioning, ever-shrinking habitats under siege from armies of hungry soldiers and poachers, reduced living space, food, and genetic vigor. Equally troubling is their silence on the crucial effects of population demographics and the rise of the agricultural state on the likelihood of human conflict.

The Napoleonic Invasion (The Yanomami Controversy)

As the summer of love was winding down and Jane Goodall's first reports of chimpanzee warfare were exploding into public consciousness, Napoleon Chagnon suddenly became the world's most famous living anthropologist with the publication of *Yanomamö: The Fierce People.* The year 1968 was a good one to come out with a dashing anthropological adventure yarn claiming to prove that warfare is ancient and integral to human nature.

The year began with the revolution in Prague and the TET offensive in Vietnam. Martin Luther King, Jr.'s worst dream came true in Memphis, Robert Kennedy was felled in a hotel kitchen, and blood and chaos ran in the streets of Chicago. Richard Nixon slinked into the White House, Charles Manson and his lost followers plotted mayhem in the dry hills above Malibu, and the Beatles put the final touches on *The White Album.* The year ended with three American astronauts, for the first time ever, gazing back upon this fragile blue planet floating in eternal silence, praying for peace.[27]

Given all that, perhaps it's not surprising that Chagnon's account of the "chronic warfare" of the "innately violent" Yanomami struck a public nerve. Desperate to understand human murderousness, the public lapped up his depictions of the day-to-day brutality of people he described as our "contemporary ancestors." Now in its fifth edition, *Yanomamö: The Fierce People* is still the all-time bestseller in anthropology, with millions of copies sold to university students alone. Chagnon's books and films have figured prominently in the education of several generations of anthropologists, most of whom accepted his claims to have demonstrated the inherent ferocity of our species.

But Chagnon's research should be approached with caution, as he employs a host of dubious techniques. Ferguson found, for example, that Chagnon conflates common murder with war in his statistics, as does Pinker in his discussion of the Gebusi. But more importantly, Chagnon fails to account for the effects of his own disruptive, rather

Hemingway-esque presence among the people he studied. According to Patrick Tierney, author of *Darkness in El Dorado*, "The wars that made Chagnon and the Yanomami famous—the ones he wrote about with such relish in *The Fierce People*—began on November 14, 1964, the same day the anthropologist arrived with his shotguns, outboard motor, and a canoe full of steel goods to give away."[28] Tierney cites Chagnon's own doctoral thesis, showing that in the thirteen years prior to his arrival, no Namowei (a large branch of the Yanomami) had been killed in warfare. But during his thirteen-month residence among them, ten Yanomami died in a conflict between the Namowei and the Patanowateri (another branch).

Kenneth Good, an anthropologist who first went to live with the Yanomami as one of Chagnon's graduate students and stayed on for twelve years, described Chagnon as "a hit-and-run anthropologist who comes into villages with armloads of machetes to purchase cooperation for his research. Unfortunately," wrote Good, "he creates conflict and division wherever he goes."[29]

Part of Chagnon's disruptiveness no doubt resulted from his blustery, macho self-conception, but his research goals may have been an even bigger source of problems. He wanted to collect genealogical information from the Yanomami. This is a tricky proposition, to say the least, given that the Yanomami consider it disrespectful to speak names out loud. Naming the dead requires breaking one of the strongest taboos in their culture. Juan Finkers, who lived among them for twenty-five years, says, "To name the dead, among the Yanomami, is a grave insult, a motive of division, fights, and wars."[30] Anthropologist Marshall Sahlins described Chagnon's research as "an absurdist anthropological project," trying to work out ancestor-based lineages "among a people who by taboo could not know, could not trace and could not name their ancestors—or for that matter, could not bear to hear their own names."[31]

Chagnon dealt with his hosts' taboo by playing one village off against another. In his own account, he:

> began taking advantage of local arguments and animosities in selecting my informants. . . . traveling to other villages to check the genealogies, picking villages that were on strained terms with the people about whom I wanted information. I would then return to my base camp and check with local informants the accuracy of the new information. If the informants became angry when I mentioned the new names I acquired from the unfriendly group, I was almost certain that the information was accurate. . . . I occasionally hit a name that put the informant into a rage, such as that of a dead brother or sister that other informants had not reported.[32]

To recap:

1. Our hero swashbuckles into Yanomami lands, bringing machetes, axes, and shotguns he presents to a few select groups, thereby creating disruptive power imbalances between groups.
2. He detects and aggravates preexisting tensions between communities by goading them to disrespect each other's honored ancestors and dead loved ones.
3. Inflaming the situation further, Chagnon reports the offenses he's provoked, using the resulting rage to confirm the validity of his genealogical data.
4. Having thus inflicted and salted the Yanomami's wounds, Chagnon sallies forth to seduce the American public with tales of derring-do among the vicious and violent "savages."

The word *anthro* has entered the vocabulary of the Yanomami. It signifies "a powerful nonhuman with deeply disturbed tendencies and wild eccentricities."[33] Since 1995, Chagnon has been legally barred from returning to the lands of the Yanomami.

When anthropologist Leslie Sponsel lived among the Yanomami in the mid-1970s, he saw no warfare, just one physical fight, and heard a few loud marital disputes. "To my surprise," writes Sponsel, "people in [my] village and three neighboring villages were simply nothing like 'the fierce people' described by Chagnon." Sponsel had brought along a copy of Chagnon's book, with its photos of fighting Yanomami warriors, as a way to explain the sort of work he was doing. "Although some of the men were absorbed by the pictures," he writes, "I was asked not to show them to children as they provided examples of undesirable behavior. These Yanomami," Sponsel concluded, "did not value fierceness in any positive way."[34]

For his part, in over a decade living among them, Good witnessed a single outbreak of war. He cut his association with Chagnon eventually, having concluded the emphasis on Yanomami violence was "contrived and distorted." Good later wrote that Chagnon's book had "blown the subject out of any sane proportion," arguing that "what he had done was tantamount to saying that New Yorkers are muggers and murderers."

The Desperate Search for Hippie Hypocrisy and Bonobo Brutality

For a certain kind of journalist (or evolutionary psychologist), nothing is more satisfying than exposing hippie hypocrisy. A recent headline from Reuters reads, "Hippie Apes Make War as Well as Love, Study Finds."[35] The article states, "Despite their reputation as lovers, not fighters, of the primate world, bonobos actually hunt and kill monkeys. . . ." Another assures us that "Despite 'Peacenik' Reputation, Bonobos Hunt and Eat Other Primates Too." A third, under the headline "Sex Crazed Apes Feast on Killing, Too," opens with an audible sneer: "As hippies had Altamont [where Hell's Angels killed a concertgoer], so bonobos have Salonga National Park, where scientists have witnessed the supposedly peace-loving primate hunting and eating monkey children." "Sex crazed"? "*Supposedly* peace-loving"? "Eating monkey children"? Do monkeys have "children"?

If both chimps *and* bonobos make war, maybe we *are* "dazed survivors" of a "5-million-year habit of lethal aggression" after all. But a closer look reveals that it's the journalists who are a bit dazed. Researchers witnessed ten attempts to hunt monkeys over five years of observing the bonobos in question. The bonobos were successful three times, sharing the monkey meat among the hunters—mixed groups of males and females.

A brief reality check for science journalists:

- Researchers have long known and reported that bonobos regularly hunt and eat meat, generally small jungle antelopes known as duikers—as well as squirrels, insects, and grubs.
- The evolutionary line leading to humans, chimps, and bonobos split from that leading to monkeys about thirty million years ago. Chimps and bonobos, in other words, are as closely related to monkeys as we are.
- Young monkeys are not "children."
- Monkey meat is on the menu at fancy Chinese restaurants and jungle barbecues in many parts of the world.
- Tens of thousands of monkeys, young and old, are sacrificed in research laboratories throughout the world annually.

So, are humans also "at war" with monkeys?

Nothing sells newspapers like headlines of "WAR!," and no doubt "CANNIBALISTIC HIPPIE ORGY WAR!" sells even more, but one species hunting and eating another species is hardly "war"; it's lunch. That bonobos and monkeys may look similar to untrained eyes is irrelevant. When a pack of wolves or coyotes attacks a stray dog, is that "war"? We've seen hawks pluck pigeons out of the sky. War?

Asking whether our species is *naturally* peaceful or warlike, generous or possessive, free-loving or jealous, is like asking whether H_2O is *naturally* a solid, liquid, or gas. The only meaningful answer to such a question is: It depends. On a nearly empty planet, with food and shelter distributed widely, avoiding conflict would have been an easy, attractive option. Under the conditions typical of ancestral environments, human beings would have had much more to lose than to gain from warring against one another. The evidence—both physical and circumstantial—points to a human prehistory in which our ancestors made far more love than war.

CHAPTER FOURTEEN

The Longevity Lie (Short?)

The days of our years are threescore years and ten;
and if by reason of strength they be fourscore years,
yet is their strength labor and sorrow;
for it is soon cut off, and we fly away.

PSALMS 90:10

Strange but true: The average height expectancy of prehistoric humans was about three feet, so a four-foot tall man was considered a giant.

Does that fact alter your image of prehistory? Are you picturing a diminutive race of bonsai people living in mini-caves, chasing rabbits into holes, cowering in fear of foxes, being carried off by hawks? Does this cause you to rethink what a challenge a mammoth hunt must have been for our half-pint ancestors? Does it make you feel even luckier to be living today, when our superior diet and sanitation have doubled the average person's height expectancy?

Well, don't get carried away. While it is *technically* true that the average "height expectancy" of prehistoric men was about three feet, it's a misleading sort of truth. Like overconfident declarations about the universality of marriage, poverty, and war, it's the sort of assertion that sows confusion and results in a harvest of misleading data.

Take the average height of a full-grown man living in prehistoric times (using skeletal remains as a guide): about six feet tall (72 inches). Then take the average size of a prehistoric infant's skeleton (let's say about 20 inches). Then extrapolate from the ratio of infant-to-adult skeletons at known archaeological burial sites and presume that in general, for every three people who lived to adulthood, seven died as infants. Thus, owing to the high rate of infant mortality, average human height in prehistory was $(3 \times 72) + (7 \times 20) \div 10 = 35.6$ inches. Roughly three feet.[1]

Absurd? Yes. Misleading? Yup. Statistically accurate? Well, kinda.

This *height* expectancy "truth" is no more absurd or misleading than what most people are led to believe about human *life* expectancy in prehistory.

Exhibit A: In an interview with *NBC Nightly News*,[2] UCSF biophysicist Jeff Lotz was discussing the prevalence of chronic back pain in the United States. The millions of people watching that night heard him explain, "*It wasn't until two or three hundred years ago that we lived past age forty-five*, so our spines really haven't evolved to the point where they can maintain this upright posture with these large gravity loads for the duration of our lives [emphasis added]."

Exhibit B: In an otherwise solid book about women in prehistory (*The Invisible Sex*), an archaeologist, an anthropologist, and an editor of one of the world's leading science magazines team up to imagine the life of a typical woman they call Ursula, living in Europe 45,000 years ago. "Life was hard," they write, "and many, especially the young and the old, died of starvation in winter and accidents of one sort or another, as well as disease. . . . Ursula [having had her first daughter at age fifteen] lived long enough to see her first granddaughter, dying at the *ripe old age of 37* [emphasis added]."[3]

Exhibit C: In a *New York Times* article,[4] James Vaupel, director of the laboratory of survival and longevity at the Max Planck Institute for Demographic Research, explains, "There is no fixed life span." Dr. Vaupel points to the increase in life expectancy from 1840 to today in countries where the figure is rising quickest and notes that this increase is "linear, absolutely linear, with no evidence of any decline or tapering off." From this, he concludes, "There's no reason that life expectancy can't continue to go up two to three years per decade."

Except that there is. At some point, all the babies who can survive to adulthood, do. Further advances will be slight.

When Does Life Begin? When Does It End?

The preceding numbers are just as fantastical as the ones we came up with for our *average height expectancy* estimate. They are, in fact, based upon the same erroneous calculation distorted by high infant mortality rates. When this factor is eliminated, we see that prehistoric humans who survived beyond childhood typically lived from sixty-six to ninety-one years, with higher levels of overall health and mobility than we find in most Western societies today.

It's a game of averages, you see. While it's true that many infants and small children died in prehistoric populations—as indicated by the larger numbers of infant skeletons in most burial sites—these skeletons tell us nothing about what constituted a "ripe old age." *Life expectancy at birth*, which is the measure generally cited, is far from an accurate measure of the *typical life span*. When you read, "At the beginning of the 20th century, life expectancy at birth was around 45 years. It has risen to about 75 thanks to the advent of antibiotics and public health measures that allow people to survive or avoid infectious diseases,"[5] keep in mind that this dramatic increase is much more a reflection of increased infant survival than of adults living longer.

In Mozambique, where one of us was born and raised, the average life expectancy at birth for a man is currently, tragically, about forty-two years. But Cacilda's father was ninety-three when he died, riding his bicycle right to the end of the road. *He* was old. A forty-year-old is not. Not even in Mozambique.

No doubt, many prehistoric infants died from disease or harsh conditions, as do the infants of other primates, human foragers, and modern Mozambiqueans. But many anthropologists agree it's likely that a large portion of infant mortality once attributed to starvation and disease probably resulted from infanticide. They argue that foraging societies limited the number of infants so they wouldn't become a burden to the group or allow overly rapid population growth to strain food supplies.

Horrific as it may be for us to contemplate, infanticide is anything but a rarity, even today. Anthropologist Nancy Scheper-Hughes studied contemporary infant deaths in northeast Brazil, where about 20 percent of infants die in their first year. She found that women consider the deaths of some children a "blessing" if the babies were lethargic and passive. Mothers told Scheper-Hughes they were "children who wanted to die, whose will to live was not sufficiently strong or developed." Scheper-Hughes found that these children received less food and medical attention than their more vigorous siblings.[6]

Joseph Birdsell, one of the world's greatest scholars of Australian Aboriginal culture, estimated that as many as half of all infants were intentionally destroyed. Various surveys of contemporary pre-industrial societies conclude that anywhere from half to three-quarters of them practice some form of direct infanticide.

Lest we start feeling too smug in our compassion and superiority, recall the foundling hospitals of Europe. The number of babies delivered to near-certain death in France rose from 40,000 in 1784 to almost 140,000 by 1822. By 1830, there were 270 revolving boxes in French foundling hospital doors specially designed to protect the anonymity of those depositing unwanted infants. Eighty to 90 percent of these children are estimated to have died within a year of arrival.

Once our ancestors began cultivating land for food, they were running on a wheel, but never fast enough. More land provides more food. And more food means more children born and fed. More children provide more help on the farm and more soldiers. But this population growth creates demand for more land, which can be won and held only through conquest and war. Put another way, the shift to agriculture was accelerated by the seemingly irrefutable belief that it's better to take strangers' land (killing them if necessary) than to allow one's own children to die of starvation.

Closer to our own time, the BBC reports that as many as 15 percent of the *reported* deaths of female infants in parts of southern India are victims of infanticide. Millions more die in China, where female infanticide is prevalent, and has been for centuries. A late nineteenth-century missionary living in China reported that of 183 sons and 175 daughters born in a typical community, 126 of the sons lived to age ten (69 percent), while only 53 of the daughters made it that far (30 percent).[7] China's one-child policy, combined with the cultural preference for sons, has only worsened the already dismal odds of survival for female infants.[8]

There are also problematic cultural assumptions lurking within demographers' calculations, in which life is assumed to begin at birth. This view is far from universal. Societies that practice infanticide don't consider newborn infants full human beings. Rituals ranging from baptism to naming ceremonies are delayed until it is determined whether or not the child will be permitted to live. If not, from this perspective, the child was never fully *alive* anyway.[9]

Is 80 the New 30?

> *Cartoon in* The New Yorker*: Two cavemen are shown chatting, one of whom is saying, "Something's just not right—our air is clean, our water is pure, we all get plenty of exercise, everything we eat is organic and free-range, and yet nobody lives past thirty."*

Statistical distortions due to infanticide are not the only source of confusion concerning prehistoric longevity. As you might imagine, it's not so easy to determine the age at death of a skeleton that's been in the ground for thousands of years. For various technical reasons, archaeologists often underestimate the age at death. For example, archaeologists estimated the ages at death of skeletons taken from mission cemeteries in California. After the estimates had been made, written records of the actual ages at death were discovered. While the archaeologists had estimated that only about 5 percent had lived to age forty-five or beyond, the documents proved that *seven times that many* (37 percent) of the people buried in these cemeteries were over forty-five years of age when they died.[10] If estimates can be so far off on skeletons just a few hundred years old, imagine the inaccuracies with remains that are tens of thousands of years old.

One of the most reliable techniques archaeologists use to estimate age at death is dental eruption. They look at how far the molars have grown out of the jawbone, which indicates roughly how old a young adult was at death. But our "wisdom teeth" stop "erupting" in our early to mid-thirties, which means that archaeologists note the age at death of skeletons beyond this point as "35+." This *doesn't* mean that thirty-five was the age of death, but that the person was *thirty-five or older*. He or she may have been anywhere from thirty-five to one hundred years old. Nobody knows.

Somewhere along the line this notation system was mistranslated in the popular press, leaving the impression that our ancient ancestors rarely made it past thirty-five. Big mistake. A wide range of data sources (including, even, the *Old Testament*) point to a typical human life span of anywhere from seventy ("three score and ten") to over ninety years.

In one study, scientists calibrated brain and body-weight ratios across different primates, arriving at an estimate of sixty-six to seventy-eight years for *Homo sapiens*.[11] These numbers bear up under observation of modern-day foragers. Among the !Kung San, Hadza, and Aché (societies in Africa and South America), a female who lived to forty-five could be expected to survive another 20, 21.3, and 22.1 years, respectively.[12] Among the !Kung San, most people who reached sixty could reasonably expect to live another ten years or so—active years of mobility and social contribution. Anthropologist Richard Lee reported that one in ten of the !Kung he encountered in his time in Botswana were over sixty years of age.[13]

As mentioned in previous chapters, it's clear that overall human health (including longevity) took a severe hit from agriculture. The typical human diet went from extreme variety and nutritional richness to just a few types of grain, possibly supplemented by occasional meat and dairy. The Aché diet, for example, includes 78 different species of mammal, 21 species of reptiles and amphibians, more than 150 species of birds, and 14 species of fish, as well as a wide range of plants.[14]

In addition to the reduced nutritional value of the agricultural diet, the diseases deadliest to our species began their dreadful rampage when human populations turned to agriculture. Conditions were perfect: high-density population centers stewing in their own filth, domesticated animals in close proximity (adding their excrement, viruses, and parasites to the mix), and extended trade routes facilitating the movement of contagious pathogens from populations with immunity to vulnerable communities.[15]

When James Larrick and his colleagues studied the still relatively isolated Waorani Indians of Ecuador, they found no evidence of hypertension, heart disease, or cancer. No anemia or common cold. No internal parasites. No sign of previous exposure to polio, pneumonia, smallpox, chicken pox, typhus, typhoid, syphilis, tuberculosis, malaria, or serum hepatitis.[16]

This is not as surprising as it may seem, given that almost all these diseases either originated in domesticated animals or depend upon high-density population for easy transmission. The deadliest infectious diseases and parasites that have plagued our species could not have spread until after the transition to agriculture.

Table 3: Deadly Diseases from Domesticated Animals[17]

HUMAN DISEASE	ANIMAL SOURCE
Measles	Cattle (rinderpest)
Tuberculosis	Cattle
Smallpox	Cattle (cowpox)
Influenza	Pigs and birds
Pertussis	Pigs and dogs
Falciparum malaria	Birds

The dramatic increases in world population that paralleled agricultural development don't indicate increased health, but increased fertility: more people living to reproduce, but lower quality of life for those who do. Even Edgerton, who repeatedly tells the longevity lie (Foragers' "lives are short—life expectancy at birth ranges between 20 and 40 years . . ."), has to agree that, somehow, foragers managed to be healthier than agriculturalists: "Agriculturalists throughout the world were always less healthy than hunters and gatherers." The urban populations of Europe, he writes, "did not match the longevity of hunter-gatherers until the mid-nineteenth or even twentieth century."[18]

That's in Europe. People living in Africa, most of Asia, and Latin America have *still* not regained the longevity typical of their ancestors and, thanks to chronic world poverty, global warming, and AIDS, it's unlikely they will for the foreseeable future.

Once pathogens mutate into human populations from domesticated animals, they quickly migrate from one community to another. For these agents of disease, the initiation of global trade was a boon. Bubonic plague took the Silk Route to Europe. Smallpox and measles stowed away on ships headed for the New World, while syphilis appears to have hitched a ride back across the Atlantic, probably on Columbus's first return voyage. Today, the Western world flutters into annual panics over avian flu scares emanating from the Far East. Ebola, SARS, flesh-eating bacteria, the H1N1 virus (swine flu), and innumerable pathogens yet to be named keep us all compulsively washing our hands.

While there were no doubt occasional outbreaks of infectious diseases in prehistory, it's unlikely they spread far, even with high levels of sexual promiscuity. It would have been nearly impossible for pathogens to take hold in widely dispersed groups of foragers with infrequent contact between groups. The conditions necessary for devastating epidemics or pandemics just didn't exist until the agricultural revolution. The claim that modern medicine and sanitation save us from infectious diseases that ravaged pre-agricultural people (something we hear often) is like arguing that seat belts and air bags protect us from car crashes that were fatal to our prehistoric ancestors.

Stressed to Death

If an infectious virus doesn't get you, a stressed-out lifestyle and unhealthy diet probably will. Cortisol, the hormone your body releases when under stress, is the strongest immunosuppressant known. In other words, nothing weakens our defenses against disease quite like stress.

Even something as seemingly unimportant as not getting enough sleep can have a dramatic effect on immunity. Sheldon Cohen and his colleagues studied the sleep habits of 153 healthy men and women for two weeks before putting them in quarantine and exposing them to rhinovirus, which causes the common cold. The less an individual slept, the more likely he or she was to come down with a cold. Those who slept less than seven hours per night were *three times* as likely to get sick.[19]

If you want to live long, sleep more and eat less. To date, the *only* demonstrably effective method for prolonging mammalian life is severe caloric reduction. When pathologist Roy Walford fed mice about half of what they wanted to eat, they lived about twice as long—the equivalent of 160 human years. They not only lived longer, but stayed fitter and smarter as well (as judged by—you guessed it—running through mazes). Follow-up studies on insects, dogs, monkeys, and humans have confirmed the benefits of going through life hungry. Intermittent fasting was associated with more than a 40 percent reduction in heart disease risk in a study of 448 people published in the *American Journal of Cardiology* reporting that "most diseases, including cancer, diabetes and even neurodegenerative illnesses, are forestalled" by caloric reduction.[20]

These studies lead to the slacker-friendly conclusion that in the ancestral environment, where our predecessors lived hand-to-mouth, a certain amount of dietary inconsistency—perhaps exacerbated by sheer laziness interrupted by regular aerobic exercise—would have been adaptive, even healthy. To put it another way, if you hunt or gather just enough low-fat food to forestall serious hunger pangs, and spend the rest of your time in low-stress activities such as telling stories by the fire, taking extended hammock-embraced naps, and playing with children, you'd be engaged in the optimal lifestyle for human longevity.[21]

Which brings us back to the eternal question asked by foragers offered the chance to join the "civilized" world and adopt farming: Why? Why work so hard when there are so many mongongo nuts in the world? Why stress over weeding the garden when there are "plenty fish, plenty fruits, and plenty birdies"?

> *We are here on Earth to fart around, and don't let anybody tell you any different.*
>
> KURT VONNEGUT, JR.

In 1902 the *New York Times* carried a report headlined "Laziness Germ Discovered." It seems one Dr. Stiles, a zoologist at the Department of Agriculture, had discovered the germ responsible for "degenerates known as *crackers* or *poor whites*" in the "Southern States." But in fact, our laziness seems less in need of explanation than our frenzied industry.

How many beavers die in dam-construction accidents? Are birds subject to sudden spells of vertigo that send them falling from the sky? How many fish drown to death? Such events are all rather infrequent we'd wager, but the toll exacted upon humans by the chronic stress many consider a normal part of human life is massive.

In Japan, there's a word for it, *karōshi* (過労死): death from overwork. Japanese police records indicate that as many as 2,200 Japanese workers committed suicide in 2008 due to overwhelming work conditions, and five times that number died from stress-induced strokes and heart attacks, according to Rengo, a labor union federation. But whether our language contains a handy term for it or not, the devastating effects of chronic stress are not limited to Japan. Heart disease, circulatory problems, digestive disorders, insomnia, depression, sexual dysfunction, and obesity—behind every one of them lurks chronic stress.

If we really did evolve in a Hobbesian ordeal of constant terror and anxiety, if our ancestors' lives truly *were* solitary, poor, nasty, brutish, and short, why, then, are we still so vulnerable to stress?[22]

Who You Calling a Starry-Eyed Romantic, Pal?

Many otherwise reasonable people seem to have a burning need to locate the roots of war deep in our primal past, to see self-sufficient foragers as *poor*, and to spread the misbegotten gospel that three or four decades was a *ripe old age* for a human being in pre-agricultural times. But this vision of our past is demonstrably false. *¿Que pasa?*

If prehistoric life *was* a perpetual struggle that ended in early death, if ours *is* a species motivated almost exclusively by self-interest, if war *is* an ancient, biologically embedded tendency, then one can soothingly argue, as Steven Pinker does, that things are getting better all the time—that, in his Panglossian view, "we are probably living in the most peaceful moment of our species' time on Earth." That would be encouraging news, indeed, which is what most audiences want to hear, after all. We all want to believe things are getting better, that our species is learning, growing, and prospering. Who refuses congratulations for having the good sense to be alive here and now?

But just as "patriotism is the conviction that your country is superior to all others because you were born in it" (G. B. Shaw), the notion that we live in our species' "most peaceful moment" is as intellectually baseless as it is emotionally comforting. Journalist Louis Menand noted how science can fulfill a conservative, essentially political function by providing "an explanation for the way things are that does not threaten the way things are." "Why," he asks rhetorically, "should someone feel unhappy or engage in anti-social behavior when that person is living in the freest and most prosperous nation on Earth? It can't be the system!"[23] What's your problem? Everything's just fine. Life's great and getting better! Less war! Longer life! New and improved human existence!

This Madison Avenue vision of the super-duper new and improved present is framed by an utterly fictional, blood-smeared Hobbesian past. Yet it's marketed to the public as the "clear-eyed realist" position, and those who question its founding assumptions risk being dismissed as delusional romantics still grieving over the death of Janis Joplin and the demise of bell-bottoms. But that "realistic" argument is riddled with misunderstood data, mistaken interpretations, and misleading calculations. A dispassionate review of the relevant science clearly demonstrates that the tens of thousands of years before the advent of agriculture, while certainly not a time of uninterrupted utopian bliss, was for the most part characterized by robust health, peace between individuals and groups, low levels of chronic stress and high levels of overall satisfaction for most of our ancestors.

Having made this argument, have we outed ourselves as card-carrying comrades in the Delusional Utopian Movement (DUM)? Is it Rousseauian fantasy to assert that prehistory was *not* an unending nightmare? That human nature leans no more toward violence, selfishness, and exploitation than toward peace, generosity, and cooperation? That most of our ancient ancestors probably experienced a sense of communal belonging few of us can imagine today? That human sexuality probably evolved and functioned as a social bonding device and a pleasurable way to avoid and neutralize conflict? Is it silly romanticism to point out that ancient humans who survived their first few years often lived as long as the richest and luckiest of us do today, even with our high-tech coronary stents, diabetes medication, and titanium hips?

No. If you think about it, the neo-Hobbesian vision is far sunnier than ours. To have concluded, as we have, that our species has an innate capacity for love and generosity *at least* equal to our taste for destruction, for peaceful cooperation as much as coordinated attack, for an open, relaxed sexuality as much as for jealous, passion-smothering possessiveness . . . to see that both these worlds were open to us, but that around ten thousand years ago a few of our ancestors wandered off the path they'd been on forever into a garden of toil, disease, and conflict where our species has been trapped ever since . . . well, this is not exactly a rose-colored view of the overall trajectory of humankind. Who are the naïve romantics here, anyway?

PART IV

Bodies in Motion

Love's mysteries in souls do grow,
But yet the body is his book.

John Donne (1572–1631)

Everybody has a story to tell. So does every *body*, and the story told by the human body is rated XXX.

Like any narrative of prehistory, ours rests on two types of evidence: circumstantial and material. We've already covered a good bit of the circumstantial evidence. As for more tangible material evidence, the song says, "What goes up must come down," but unfortunately for archaeologists and those of us who rely on their findings, what goes down rarely comes back up. And even when it does, ancient social behavior is hard to see reflected in bits of bone, flint, and pottery—fragments that represent only a fraction of what once existed.

At a conference not long ago, the subject of our research came up over breakfast. Upon hearing that we were investigating human sexual behavior in prehistory, the professor sitting across the table from us scoffed and asked (rhetorically), "So what do you do, close your eyes and dream?" While one should never scoff with a mouthful of scone, he had a point. As social behavior presumably doesn't leave physical artifacts, any theorizing must amount to little but "dreaming."

Paleontologist Stephen Jay Gould was an early scoffer at the notion of evolutionary psychology, asking, "How can we possibly know in detail what small bands of hunter-gatherers did in Africa two million years ago?"[1] Richard Potts, director of the Smithsonian's Human Origins Program, agrees, warning, "Many characteristics of early human behaviour are . . . difficult to reconstruct, as no appropriate material evidence is available. Mating patterns and language are obvious examples . . . [they] leave no traces in the fossil record." But he then adds, as if under his breath, "Questions of social life . . . may be accessible from studies of ancient environments, or from certain aspects of anatomy and behaviour that leave material evidence."[2]

Certain aspects of anatomy and behaviour that leave material evidence. . . . Can we glean reliable information about the contours of ancient social life—even sexual behavior—from present-day human anatomy?

Yes we can.

CHAPTER FIFTEEN

Little Big Man

Every creature's body tells a detailed story about the environment in which its ancestors evolved. Its fur, fat, and feathers suggest the temperatures of ancient environments. Its teeth and digestive system contain information about primordial diet. Its eyes, legs, and feet show how its ancestors got around. The relative sizes of males and females and the particulars of their genitalia say a lot about reproduction. In fact, male sexual ornaments (such as peacock's tails or lions' manes) and genitals offer the best way to differentiate between closely related species. Evolutionary psychologist Geoffrey F. Miller goes so far as to say that "evolutionary innovation seems focused on the details of penis shape."[1]

Leaving aside for the moment the disturbingly Freudian notion that even Mother Nature is obsessed with the penis, our bodies certainly contain a wealth of information about the sexual behavior of our species over the millennia. There are clues encoded in skeletal remains millions of years old and pulsing in our own living bodies. It's all right there—and here. Rather than closing our eyes and dreaming, let's open them and learn to read the hieroglyphics of the sexual body.

We begin with body-size dimorphism. This technical-sounding term simply refers to the average difference in size between adult males and females in a given species. Among apes for example, male gorillas and orangutans average about twice the size of females, while male chimps, bonobos, and humans are from 10 to 20 percent bigger and heavier than females. Male and female gibbons are of equal stature.

Among mammals generally and particularly among primates, body-size dimorphism is correlated with male competition over mating.[2] In winner-take-all mating systems where males compete with each other over infrequent mating opportunities, the larger, stronger males tend to win . . . and take all. The biggest, baddest gorillas, for example, will pass genes for bigness and badness into the next generation, thus leading to ever bigger, badder male gorillas—until the increased size eventually runs into another factor limiting this growth.

On the other hand, in species with little struggle over females, there is less biological imperative for the males to evolve larger, stronger bodies, so they generally don't. That's why the sexually monogamous gibbons are virtually identical in size.

Looking at our modest body-size dimorphism, it's a good bet that males haven't been fighting much over females in the past few million years. As mentioned above, men's bodies are from 10 to 20 percent bigger and heavier than women's on average, a ratio that appears to have held steady for at least several million years.[3]

Owen Lovejoy has long argued that this ratio is evidence of the ancient origins of monogamy. In an article he published in *Science* in 1981, Lovejoy argued that both the accelerated brain development of our ancestors and their use of tools resulted from an "already established hominid character system," that featured "intensified parenting and social relationships, monogamous pair bonding, specialized sexual-reproductive behavior, and bipedality." Thus, Lovejoy argued, "The nuclear family and human sexual behavior may have their ultimate origin long before the dawn of the Pleistocene." In fact, he concluded with a flourish, the "unique sexual and reproductive behavior of man may be the sine qua non of human origin." Almost three decades later, Lovejoy is still pushing the same argument as this book goes to press. He argues—again in *Science*—that *Ardipithecus ramidus*' fragmentary skeletal and dental remains dated to 4.4 million years ago reinforce this view of pair bonding as *the* defining human characteristic—predating, even, our uniquely large neocortex.[4]

Like many theorists, Matt Ridley agrees with this ancient origin of monogamy, writing, "Long pair-bonds shackled each ape-man to its mate for much of its reproductive life."

Four million years is an awful lot of monogamy. Shouldn't these "shackles" be more comfortable by now?

Without access to the skeletal data on body-size dimorphism we have today, Darwin speculated that early humans may have lived in a harem-based system. But we now know that if Darwin's conjecture were correct, contemporary men would be twice the size of women, on average. And, as we'll discuss in the next section, another sure sign of a gorilla-like human past would be an embarrassing case of genital shrinkage.

Still, some continue to insist that humans are naturally polygynous harem-builders, despite the paucity of evidence supporting this argument. For example, Alan S. Miller and Satoshi Kanazawa claim that, "We know that humans have been polygynous throughout most of history because men are taller than women." These authors go on to conclude that because "human males are 10 percent taller and 20 percent heavier than females, this suggests that, throughout history, humans have been mildly polygynous."[5]

Their analysis ignores the fact that the cultural conditions necessary for some males to accumulate sufficient political power and wealth to support multiple wives and their children *simply did not exist before agriculture.* And males being moderately taller and heavier than females indicates reduced competition among males, but not necessarily "mild polygyny." After all, those promiscuous cousins of ours, chimps and bonobos, reflect *precisely the same range of male/female size difference while shamelessly enjoying uncounted sexual encounters with as many partners as they can drum up.* No one claims the 10 to 20 percent body-size dimorphism seen in chimps and bonobos is evidence of "mild polygyny." The assertion that the *same* physical evidence correlates to promiscuity in chimps and bonobos but indicates mild polygyny or monogamy in humans shows just how shaky the standard model really is.

For various reasons, prehistoric harems were unlikely for our species. The famed sexual appetites of Ismail the Bloodthirsty, Genghis Khan, Brigham Young, and Wilt Chamberlain notwithstanding, our bodies argue strongly against it. Harems result from the common male hunger for sexual variety and the post-agricultural concentration of power in the hands of a few men combined with low levels of female autonomy typical of agricultural societies. Harems are a feature of militaristic, rigidly hierarchical agricultural and pastoral cultures oriented toward rapid population growth, territorial expansion, and accumulation of wealth. Captive harems have never been reported in any immediate-return foraging society.

While our species' shift to moderate body-size dimorphism strongly suggests that males found an alternative to fighting over mating opportunities millions of years ago, it doesn't tell us what that alternative was. Many theorists have interpreted the shift as confirmation of a transition from polygyny to monogamy—but that conclusion requires us to ignore multimale–multifemale mating as an option for our ancestors. Yes, a one-man/one-woman system reduces competition among males, as the pool of available females isn't being dominated by just a few men, leaving more women available for less desired men. But a mating system in which both males and females typically have multiple sexual relationships running in parallel reduces male mating competition just as effectively, if not more so. And given that both of the species closest to us practice multimale–multifemale mating, this seems by far the more likely scenario.

Why are scientists so reluctant to consider the implications of our two closest primate relatives displaying the same levels of body-size dimorphism we do? Could it be because neither is remotely monogamous? The only two "acceptable" interpretations of this shift in body-size dimorphism appear to be:

1. It indicates the origins of our nuclear family/sexually monogamous mating system. (Then why aren't men and women the same size, like gibbons?)
2. It shows that humans are naturally polygynous but have learned to control the impulse, with mixed success. (Then why aren't men twice the size of women, like gorillas?)

Note the assumption shared by both these interpretations: female sexual reticence. In both scenarios, female "honor" is intact. In the second interpretation, only the male's natural fidelity is in doubt.

When the three most closely related apes exhibit the same degree of body-size dimorphism, shouldn't we at least consider the possibility that their bodies reflect the same adaptations before we reach for far-fetched, if emotionally reassuring, conclusions?

It's time to go below the belt. . . .

All's Fair in Love and Sperm War

No case interested and perplexed me so much as the brightly colored hinder ends and adjoining parts of certain monkeys.

CHARLES DARWIN[6]

It seems men weren't fighting much over dates for the past few million years (until agriculture), but that doesn't mean Darwin was wrong about male sexual competition being of crucial importance in human evolution. Even among bonobos, who experience little to no overt conflict over sex, Darwinian selection takes place, but on a level Darwin himself probably never considered—or dared discuss publicly, anyway. Rather than male bonobos competing to see who gets lucky, *they all get lucky*, and then let their spermatozoa fight it out. Øjvind Winge, who was working with guppies in the 1930s, coined the term *sperm competition*. Geoffrey Parker, studying the decidedly unglamorous yellow dungfly, later refined the concept.

The idea is simple. If the sperm of more than one male are present in the reproductive tract of an ovulating female, the spermatozoa themselves compete to fertilize the ovum. Females of species that engage in sperm competition typically have various tricks to advertise their fertility, thereby inviting more competitors. Their provocations range from sexy vocalizations or scents to genital swellings that turn every shade of lipstick red from Berry Sexy to Rouge Soleil.[7]

The process is something of a lottery, where the male with the most tickets has the best shot at winning (hence, the chimp and bonobo's huge sperm-production capabilities). It's also an obstacle course, with the female's body providing various types of hoops to jump through and moats to swim across to reach the egg—thus eliminating unworthy sperm. (We'll examine some of these obstacles in following chapters.) Some researchers argue that the competition is more like rugby, with various sperm forming "teams" with specialized blockers, runners, and so on.[8] Sperm competition takes many forms.

Although Darwin might have been "perplexed" by it, sperm competition preserves the central purpose of male competition in his theory of sexual selection, with the reward to the victor being fertilization of the ovum. But the struggle occurs on the cellular level, among the sperm cells, with the female's reproductive tract the field of battle. Male apes living in multimale social groups (such as chimps, bonobos, and humans) have larger testes, housed in an external scrotum, mature later than females, and produce larger volumes of ejaculate containing greater concentrations of sperm cells than primates in which females normally mate with only one male per cycle (such as gorillas, gibbons, and orangutans).

And, who knows? Perhaps Darwin would have recognized this process, if he'd been a bit less indoctrinated by Victorian notions of female sexuality. Sarah Hrdy contends that "it was Darwin's presumption that females hold themselves in reserve for the best available male that left him so puzzled by sexual swellings." Hrdy doesn't buy Darwin's "coy female" schtick for a minute: "Although appropriate for many animals, the appellation 'coy'—which was to remain unchallenged dogma for the succeeding hundred years—did not then, and does not today, apply to the observed behavior of monkey and ape females at mid-cycle."[9]

It's possible Darwin was being a bit coy himself in his writings on the sexuality of human females. The poor guy had already insulted God, as most people—including his loving, pious wife—understood the concept. Even if he suspected something like sperm competition *had* played a role in human evolution, Darwin could hardly be expected to drag the angelic Victorian woman down from her pedestal. Bad enough that Darwinian theory depicts women evolving to prostitute themselves for meat, access to male wealth, and the rest of it. To have argued that ancestral females were shameless trollops motivated by erotic pleasure would have been too much.

Still, with characteristic awareness of just how much he didn't—and couldn't—know, Darwin acknowledged, "As these parts are more brightly coloured in one sex [female] than in the other, and as they become more brilliant during the season of love, I concluded that the colours had been gained as a sexual attraction. I was well aware that I thus laid myself open to ridicule. . . ."[10]

Perhaps Darwin *did* understand that the bright sexual swellings of some female primates served to stoke male libido, which shouldn't be necessary according to his sexual selection theory. There is even evidence that Darwin may have had reason to ponder sperm competition in humans. In a letter from Bhutan, where he was gathering plants,

Darwin's old friend Joseph Hooker discussed the polyandrous humans he was encountering, where "a wife may have 10 husbands by law."

Moderate body-size dimorphism isn't the only anatomical suggestion of promiscuity in our species. The ratio of testicular volume to overall body mass can be used to read the degree of sperm competition in any given species. Jared Diamond considers the theory of testis size to be "one of the triumphs of modern physical anthropology."[11] Like most great ideas, the theory of testis size is simple: species that copulate more often need larger testes, and species in which several males routinely copulate with one ovulating female need even bigger testes.

If a species has *cojones grandes*, you can bet that males have frequent ejaculations with females who sleep around. Where the females save it for Mr. Right, the males have smaller testes, relative to their overall body mass. The correlation of slutty females with big balled males appears to apply not only to humans and other primates, but to many mammals, as well as to birds, butterflies, reptiles, and fish.

In gorillas' winner-take-all approach to mating, males compete to see who gets *all* the booty, as it were. So, although an adult silverback gorilla weighs in at around four hundred pounds, his penis is just over an inch long, at full mast, and his testicles are the size of kidney beans, though you'd have trouble finding them, as they're safely tucked up inside his body. A one-hundred-pound bonobo has a penis three times as long as the gorilla's and testicles the size of chicken eggs. The extra-large, AAA type (see chart on page 224). In bonobos, since everybody gets some sugar, the competition takes place on the level of the sperm cell, not at the level of the individual male. Still, although almost all bonobos are having sex, given the realities of biological reproduction, each baby bonobo still has only one biological father.

So the game's still the same—getting one's genes into the future—but the field of play is different. With harem-based polygynous systems like the gorilla's, individual males fight it out before any sex takes place. In sperm competition, the cells fight *in there* so males don't have to fight *out here*. Instead, males can relax around one another, allowing larger group sizes, enhancing cooperation, and avoiding disruption to the social dynamic. This helps explain why no primate living in multimale social groups is monogamous. It just wouldn't work.

As always, natural selection targets the relevant organs and systems for adaptation. Through the generations, male gorillas evolved impressive muscles for their reproductive struggle, while their relatively unimportant genitals dwindled down to the bare minimum needed for uncontested fertilization. Conversely, male chimps, bonobos, and humans had less need for oversized muscles for fighting but evolved larger, more powerful testicles and, in the case of humans, a much more interesting penis.

We can almost hear some of our readers thinking, "But my testicles aren't the size of chicken eggs!" No, they're not. But we're guessing they're not tiny kidney beans tucked up inside your abdomen, either. Humans fall in the middle ground between gorillas and bonobos on the testicular volume/body-mass scale. Those who argue that our species has been sexually monogamous for millions of years point out that human testicles are smaller than those of chimps and bonobos. Those who challenge the standard narrative (like us, for example) note that human testicular ratios are far beyond those of the polygynous gorilla or the monogamous gibbon.

So, is the human scrotum half-empty or half-full?

Multiple Comparisons of Ape Anatomy

Information Indicated:

- M/F body-size dimorphism (av. weight);
- Coitus typically face-to-face or rear entry;
- Testicular volume/body mass;
- Testicles inside body or in external scrotum;
- Comparative penis length (erect);
- Presence of pendulous breasts;
- Swelling of female genitalia at ovulation.

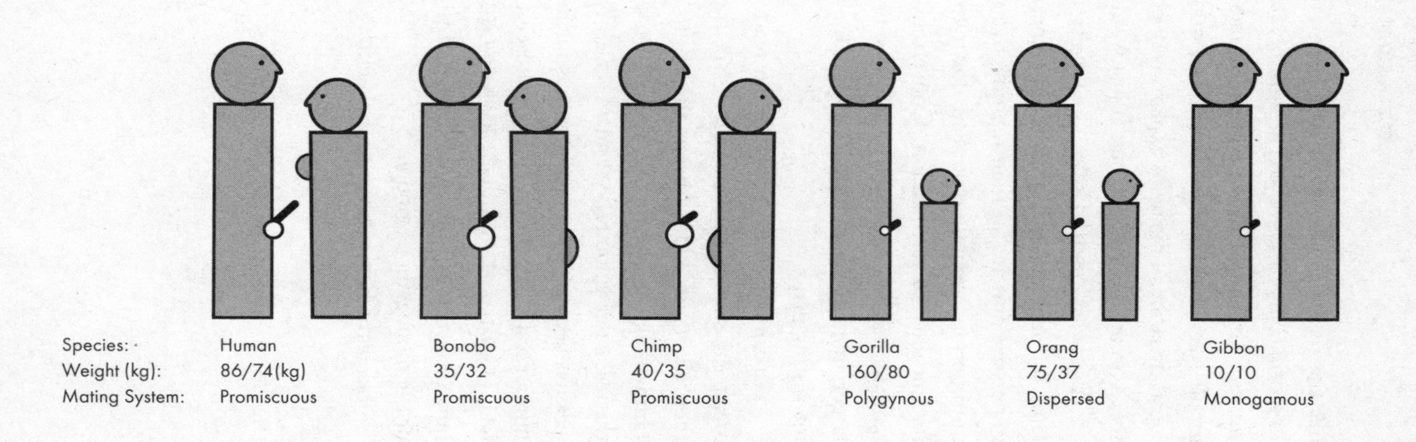

Species:	Human	Bonobo	Chimp	Gorilla	Orang	Gibbon
Weight (kg):	86/74(kg)	35/32	40/35	160/80	75/37	10/10
Mating System:	Promiscuous	Promiscuous	Promiscuous	Polygynous	Dispersed	Monogamous

CHAPTER SIXTEEN

The Truest Measure of a Man

Small?

Both chimps and bonobos are far more promiscuous than we are. Our testicles reflect this: they are mere peanuts compared to our ape relatives' coconuts.

FRANS DE WAAL[1]

Medium?

Convincing vestiges of a sexual selective history in which females mated polyandrously can be found in the human male. Perhaps the clearest vestige is testis size. Men's testes are substantially larger, relative to body size, than those of gorillas.

MARGO WILSON AND MARTIN DALY[2]

Large?

Human beings are definitely at the big-ball end of the primate spectrum, more like chimpanzees than like gorillas . . . suggesting that we have long been accustomed to competing via our sperm as well as our bodies.

DAVID BARASH AND JUDITH LIPTON[3]

As you can see, there is fundamental disagreement over the male package. What are we talking about here? Peanuts or walnuts? Ping-pong or bowling? Modern men's testicles are smaller than those of chimps and bonobos, yet they put those of polygynous gorillas and monogamous gibbons to shame, tipping the scales at about half an ounce each (that's about eighty carats, if you're a jeweler). Thus both sides of this pivotal debate can claim evidence for their view simply by declaring that human testicles are relatively large or relatively small.

But measuring a testicle is not quite like getting a shoe size. The argument that modern men's testicles would be as big as the chimps' if we'd evolved in promiscuous groups is founded on a crucial and erroneous assumption: that human testicles haven't changed in ten thousand years. When Stephen Jay Gould wrote, "There's been no biological change in humans in 40,000 or 50,000 years," he was relying on data that have been supplanted since he died in 2002. This still widely shared assumption grows out of the long-standing belief that evolution operates extremely slowly, requiring thousands of generations to make significant changes.

Sometimes it does. But sometimes it doesn't. In *The 10,000 Year Explosion*, Gregory Cochran and Henry Harpending show that the human body is capable of very rapid evolutionary change. "Humans have changed significantly in body and mind over recorded history," they write, citing resistance to malaria, blue eye color, and lactose tolerance as examples of accelerated evolutionary change since the advent of agriculture.

An example they don't discuss in their book, but might consider for future editions, is testicular size. Testicles can change size almost in the blink of an eye—blue or otherwise. In some species of lemurs (small, nocturnal primates), testicular volume changes seasonally, swelling up in breeding season and then shrinking back down in the off-season, like a beach ball with a slow leak.[4]

Testicular tissue in humans, chimps, and bonobos (but, interestingly, not gorillas) is controlled by DNA that responds unusually rapidly to environmental changes. Writing in *Nature*, geneticists Gerald Wyckoff, Hurng-Yi Wang, and Chung-I Wu report, "Rapid evolution of male reproductive genes is . . . quite notable in the lineages to human and chimpanzee." They go on to note that the rapid response of these genes may well be associated with mating systems: "The contrast is intriguing in light of the socio-sexual behaviors of African apes. Whereas modern chimpanzees and bonobos are clearly promiscuous with ample chance of multiple insemination, ovulating female gorillas seem much less likely to be multiply inseminated."[5]

Let that soak in for a moment. Humans, chimps, and bonobos—but *not* gorillas—show "accelerated evolution of genes involved in sperm and seminal fluid production" associated with "multiple insemination." The genes associated with the development of human, chimp, and bonobo testicles are highly responsive to adaptive pressures, far more so than these same genes in gorillas, where females typically mate with only one male.

Because they are composed entirely of soft tissue, testicles don't leave a trace in fossils. So, while defenders of the standard narrative assume that human testicular volume has remained constant for millennia, it's now clear this assumption may be wrong.

Wyckoff, Wang, and Wu confirm a prediction made by biologist Roger Short back in 1979, when he wrote, "Testis size might be expected to respond rapidly to selection pressures. One of the most intense forms of selection will be found in promiscuous mating systems. . . ."[6] Geoffrey Miller agrees: "Heritable differences in sperm quality and sperm delivery equipment will be under intense selection." Finally, evolutionary biologist Lynn Margulis and her co-author Dorion Sagan reason that men's "souped-up genitals" backed by "a lot of spermatic firepower" would be worthwhile only if there were "some sort of race or contest. Otherwise," they write, "they seem excessive."[7]

Souped-up genitals. Spermatic firepower. Now we're talking!

Indications of spermatic firepower are evident in the differences between a man's first spurts and his last. A human ejaculation typically consists of anywhere from three to nine spurts. Researchers who somehow managed to capture "split ejaculates" for analysis found that the first spurts contain chemicals that protect against various kinds of chemical attack. What sort of chemical attack? Aside from leucocytes and antigens present in a woman's reproductive tract (more on that later), they protect the sperm from the chemicals in the latter spurts of other men's ejaculate. These final spurts contain a spermicidal substance that slows the advance of any latecomers. In other words, competing sperm from other men seems to be anticipated in the chemistry of men's semen, both in the early spurts (protective) and in the later spurts (attacking).[8]

The importance of sperm competition has been debated in scientific conferences and academic journals for the past few decades as if it were a new discovery, but several centuries BCE, Aristotle and his predecessors noted that if a bitch copulated with two dogs during a single fertile period, she could produce a litter of pups fathered by one or both of them. And consider the story of Heracles and Iphicles: the night preceding Amphitryon's marriage to Alcmene, Zeus disguised himself as Amphitryon and slept with the bride-to-be. The following night, Amphitryon consummated her marriage. Alcmene had twins: Iphicles (fathered by Amphitryon) and Heracles (fathered by Zeus). Clearly, the ancient Greeks had an inkling of sperm competition.

More recently, several researchers have demonstrated that a man's sperm production increases significantly when he has not seen his partner for a few days, regardless of whether or not he ejaculated during her absence. This finding conforms to the notion that sperm competition has played a role in human evolution and may even reflect an adaptation to monogamy. In this scenario, not knowing what his strumpet of a wife was up to at that damned conference in Orlando leads a man's body to hyper-produce sperm to increase his chances of fertilizing her ovum when she gets home, even if his worst fears (and possibly, hottest fantasies) are true. Along these same lines, women have also reported that their partners tend to be more vigorous in bed—reporting deeper, more vigorous thrusting—after a separation or if infidelity is suspected.[9] (The possibility that the men may actually be *turned on* by the thought of their mates' possible transgressions appears not to have made its way into the discussion as yet, but see the discussion of porn below.)

The scandalous implications of sperm competition run smack into the long-held view of sacrosanct female sexuality. It's a vision Darwin cultivated in public consciousness, featuring coy females who surrender only to a carefully chosen mate who has proven himself worthy—and even then, she's only doing it for England. "The sexually insatiable woman," declared a terrified Donald Symons, "is to be found primarily, if not exclusively, in the ideology of feminism, the hopes of boys, and the fears of men."[10] Perhaps, but Marvin Harris offers a different take, writing, "Like all dominant groups, men seek to promote an image of their subordinate's nature that contributes to the preservation of the status quo. For thousands of years, males have seen women not as women could be, but only as males want them to be."[11]

Despite all the controversy, there is no question as to whether sperm competition occurs in human reproduction.[12] It does—every time. A single human ejaculate contains anywhere from fifty million to half a billion applicants all trying to elbow their way into the only job available: fertilizer in chief. The relevant question is whether those applicants are competing against only each other, or billions more eager job-seekers sent by other men as well.

It's hard to conjure a more purely competitive entity than the human spermatozoa. Imagine a teeming school of microscopic salmon whose entire existence consists in straining upstream toward a one-in-several-hundred-million shot at reproduction. Longish odds, you might say. But not all creatures' sperm face such overwhelmingly slim chances. In some species of insects, for example, fewer than a hundred sperm line up for the race to the ovum. Nor are all sperm tiny in comparison to him that sent them. Some species of fruit fly have sperm that measure almost six centimeters when uncoiled—several times larger than the fly himself. *Homo sapiens* is far down at the other extreme, depositing hundreds of millions of tiny sperm at the drop of a hat.

Table 4: Sperm Competition Among Great Apes[13]

Ape	Human	Chimp/Bonobo*	Orang	Gorilla
Body-Size Dimorphism (%)	15–20	15–20	100	100
Testes Mass (Combined, Absolute, grams)	35–50	118–160	35	29
Seminal Volume per Ejaculate (mL)	4.25 (2–6.5)	1.1	1.1	0.3
Sperm Concentration ($\times 10^6$/mL)	1940: 113 1990: 66	548	61	171
Total Sperm Count (Million Sperm/Ejaculate)	1940: 480 1990: 280	603	67	51
Seminal Vesicles	Medium	Large	Large	Small
Penis Thickness (Circumference)	24.5mm	12mm	n/a	n/a
Penis Length	13–18cm**	7.5cm	4cm	3cm
Penis Length (Relative to Body Mass)	0.163	0.195	0.053	0.018
Body Mass (Male, kg)	77	46	45–100	136–204
Copulations per Birth (Approx.)	>1,000	>1,000	<20	<20
Average Copulation Duration (seconds)	474	7/15	900	60

* No significant differences between them in these areas.

** The coronal ridge and glans are unique to humans among apes.

Hard Core in the Stone Age

Here's a head-scratcher. Why do so many heterosexual men get off on pornography featuring groups of guys having sex with just one woman? It doesn't add up, if you think about it. It's more cone than ice cream. And the suggestive strangeness doesn't end with the counterintuitive male-to-female ratio; it's male ejaculate that puts the money in the *money shot*.

Researchers have confirmed what porn producers already know: men tend to get turned on by images depicting an environment in which sperm competition is clearly at play (though few, we imagine, think of it in quite these terms). Images and videos showing one woman with multiple males are far more popular on the Internet and in commercial pornography than those depicting one male with multiple females.[14] A quick peek at the online offerings at Adult Video Universe lists over nine hundred titles in the *Gangbang* genre, but only twenty-seven listed under *Reverse Gangbang*. You do the math. Why would the males in a species that's been *wearing the shackles of monogamy for 1.9 million years* be sexually excited by scenes of groups of men ejaculating with one or two women?

Skeptics may argue that this arousal could reflect nothing more than commercial interests or a passing fashion. Fair enough, but what to make of experimental evidence that men viewing erotic material suggestive of sperm competition (two men with one woman) produce ejaculates containing a higher percentage of motile sperm than men viewing explicit images of only three women?[15] And why does being cuckolded consistently appear at or near the top of married men's sexual fantasies, according to experts ranging from Alfred Kinsey to Dan Savage?

As far as we know, there is no corresponding taste among women for erotica featuring multiple overweight middle-aged ladies with cheap tattoos, bad haircuts, and black socks having sex with one hot guy. Go figure.

Could this male appetite for multimale scenes be an echo of the *porn of the Pleistocene*? Keep in mind the variety of societies discussed previously in which women assist and inspire teams of workers or hunters by making themselves available for sequential sex. The same dynamic is hinted at on any given Sunday with fluttering pom-poms, the shortest of shorts, and the highest of kicks ending with sexy young legs spread right down to the Astroturf. While there are other conceivable explanations for such oddities of contemporary life, they certainly align well with a prehistory characterized by sperm competition.[16] Go Trojans!

CHAPTER SEVENTEEN

Sometimes a Penis Is Just a Penis

We are right to note the license and disobedience of this member which thrusts itself forward too inopportunely when we do not want it to, and which so inopportunely lets us down when we most need it; it imperiously contests for authority with our will: it stubbornly and proudly refuses all our incitements, both of the mind and hand.

MICHEL MONTAIGNE on the penis (his, presumably)

Don't be distracted by the snickering. The human male takes his genitalia very seriously. In ancient Rome, rich boys wore a *bulla*: a locket holding a replica of a tiny hard-on. This rocket-in-a-locket was known as a *fascinum*, and signified the young man's upper-class status. "Today," writes David Friedman in *A Mind of Its Own*, his amusing and erudite history of the penis, "fifteen hundred years after the fall of Imperial Rome, anything as powerful or intriguing as an erection is said to be 'fascinating.'" Going back a bit farther, we find that in the biblical books of Genesis and Exodus, Jacob's children sprang from his *thigh*. Most historians agree that "thigh" is actually a polite way of referring to that which hangs between a man's thighs. "It seems clear," writes

Friedman, "that sacred oaths between Israelites were sealed by placing a hand on the male member." So, according to Friedman at least, the act of swearing on one's balls lives on in the world today.

Historical oddities aside, some argue that moderately sized human testicles and lower human sperm concentration (relative to chimpanzees and bonobos) disprove any significant sperm competition in human evolution. True, the reported range of human sperm concentration of $60–235 \times 10^6$ per mL pales in comparison to that of chimps, an impressive 548×10^6. But not all sperm competition is created equal.

For example, some species have seminal fluid that forms a "copulatory plug," serving to block the entrance of any subsequent sperm into the cervical canal. Species engaging in this type of sperm competition (snakes, rodents, some insects, kangaroos) typically wield penises with elaborate hooks or curlicues on the end that function to pull any previous male's plug out of the cervical opening. Though at least one team of researchers reports data suggesting men who copulate frequently produce semen that coagulates for a longer time, copulatory plugs don't appear to be in the human sexual arsenal.

Despite its lack of curlicues, the human penis is not without interesting design features. Primate sexuality expert Alan Dixson writes, "In primates which live in family groups consisting of an adult pair plus offspring [such as gibbons] the male usually has a small and relatively unspecialized penis." Say what you will about the human penis, but it ain't small or unspecialized. Reproductive biologist Roger Short (real name) writes, "The great size of the erect human penis, in marked contrast to that of the Great Apes, makes one wonder what particular evolutionary forces have been at work." Geoffrey Miller just comes out and says it: "Adult male humans have the longest, thickest, and most flexible penises of any living primate."[1] So there.

Homo sapiens: the great ape with the great penis!

The unusual flared glans of the human penis forming the coronal ridge, combined with the repeated thrusting action characteristic of human intercourse—ranging anywhere from ten to five hundred thrusts per romantic interlude—creates a vacuum in the female's reproductive tract. This vacuum pulls any previously deposited semen away from the ovum, thus aiding the sperm about to be sent into action. But wouldn't this vacuum action also draw away a man's own sperm? No, because upon ejaculation, the head of the penis shrinks in size before any loss of tumescence (stiffness) in the shaft, thus neutralizing the suction that might have pulled his own boys back.[2] Very clever.

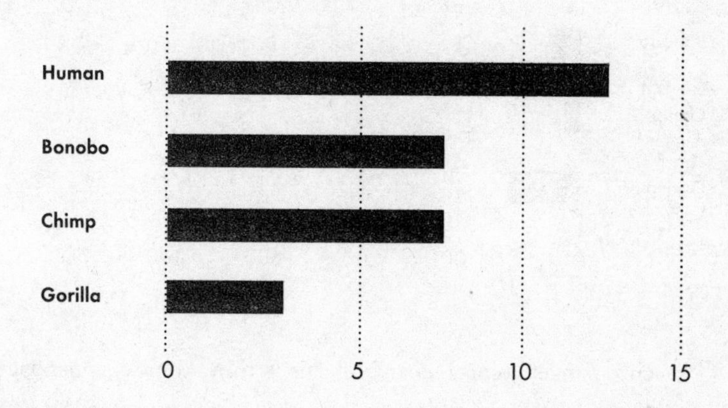

Intrepid researchers have demonstrated this process, known as semen displacement, using artificial semen made of cornstarch (the same recipe used to simulate exaggerated ejaculates in many pornographic films), latex vaginas, and artificial penises in a proper university laboratory setting. Professor Gordon G. Gallup and his team reported that more than 90 percent of the cornstarch mixture was displaced with just *a single thrust* of their lab penis. "We theorize that as a consequence of competition for paternity, human males evolved uniquely configured penises that function to displace semen from the vagina left by other males," Gallup told *BBC News Online*.

It bears repeating that the human penis is the longest and thickest of any primate's—in both absolute and relative terms. And despite all the bad press they get, men last far longer in the saddle than bonobos (fifteen seconds), chimps (seven seconds), or gorillas (sixty seconds), clocking in between four and seven minutes, on average.

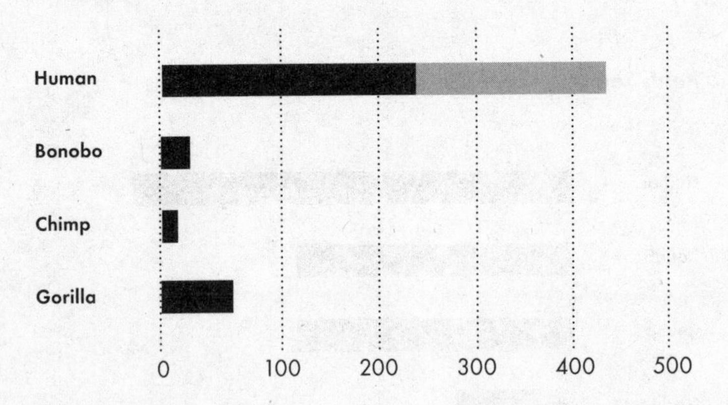

The chimpanzee penis, meanwhile, is a thin, conical appendage without the flared glans of the human member. Nor is sustained thrusting common to chimpanzee or bonobo copulation. (But really, how much sustained *anything* can you expect in seven seconds?) So while our closest ape cousins may have us beat in the testicles department, they lose to the human penis on size, duration, and cool design features. Furthermore, the average seminal volume in a human ejaculate is about four times that of chimpanzees, bringing the total number of sperm cells per ejaculate within range of the chimp's.

Returning to the question of whether the human scrotum is half-empty or half-full, the very existence of the external human scrotum suggests sperm competition in human evolution. Gorillas and gibbons, like most other mammals that don't engage in sperm competition, generally aren't equipped for it.[3]

A scrotum is like a spare refrigerator in the garage just for beer. If you've got a spare beer fridge, you're probably the type who expects a party to break out at any moment. You want to be prepared. A scrotum fulfills the same function. By keeping the testicles a few degrees cooler than they would be inside the body, a scrotum allows chilled spermatozoa to accumulate and remain viable longer, available if needed.

Anyone who's been kicked in the beer fridge can tell you this is a potentially costly arrangement. The increased vulnerability of having testes out there in the wind inviting attack or accident rather than tucked away safely inside the body is hard to overstate—especially if you're crumpled in the fetal position, unable to breathe. Given the unrelenting logic of evolutionary cost/benefit analyses, we can be quite certain this is not an adaptation without good reason.[4] Why carry the tools if you don't have the job?

There is compelling evidence pointing to a dramatic reduction in human sperm production and testicular volume in recent times. Researchers have documented worrisome decreases in average sperm count as well as reduced vitality of the sperm that do survive. One researcher suggests that average sperm counts in Danish men have plummeted from 113×10^6 in 1940 to about half that in 1990 (66×10^6).[5] The list of potential causes for the collapse is long, ranging from estrogen-like compounds in soybeans and the milk of pregnant cows to pesticides, fertilizers, growth hormones in cattle, and chemicals used in plastics. Recent research suggests the widely prescribed antidepressant paroxetine (sold as Seroxat or Paxil) may damage DNA in sperm cells.[6] The Human Reproduction study at the University of Rochester found that men whose mothers had eaten beef more than seven times per week while pregnant were three times more likely to be classified as subfertile (fewer than twenty million sperm per milliliter of seminal fluid). Among these sons of beef eaters, the rate of subfertility was 17.7 percent, as opposed to 5.7 percent among men whose mothers ate beef less often.

Humans seem to have much more sperm-producing tissue than any monogamous or polygynous primate would need. Men produce only about one-third to one-eighth as much sperm per gram of spermatogenic tissue as eight other mammals tested.[7] Researchers have noted similar surplus capacities in other aspects of human sperm and semen-production physiology.[8]

The correlation between infrequent ejaculation and various health problems offers further evidence that present-day men are not using their reproductive equipment to its fullest potential. A team of Australian researchers, for example, found that men who had ejaculated more than five times per week between the ages of twenty and fifty were one-third less likely to develop prostate cancer later in life.[9] Along with the fructose, potassium, zinc, and other benign components of semen, trace amounts of carcinogens are often present, so researchers hypothesize that the reduction in cancer rates may be due to the frequent flushing of the ducts.

A different team from Sydney University reported in late 2007 that daily ejaculation dramatically reduced DNA damage to men's sperm cells, thereby *increasing* male fertility—quite the opposite of the conventional wisdom. After forty-two men with damaged sperm were instructed to ejaculate daily for a week, almost all showed less chromosomal damage than a control group who had abstained for three days.[10]

Frequent orgasm is associated with better cardiac health as well. A study conducted at the University of Bristol and Queen's University of Belfast found that men who have three or more orgasms per week are 50 percent less likely to die from coronary heart disease.[11]

Use it or lose it is one of the basic tenets of natural selection. With its relentless economizing, evolution rarely equips an organism for a task not performed. If contemporary levels of sperm and semen production were typical of our ancestors, it is unlikely our species would have evolved so much surplus capacity. Contemporary men have far more potential than they use. But if it's true that modern human testicles are mere shadows of their former selves, what happened?

Since the infertile leave no descendants, it's a truism in evolutionary theory that infertility can't be inherited. But *low fertility* can be passed along under certain conditions. As discussed above, chromosomes in humans, chimps, and bonobos associated with sperm-producing tissue respond very rapidly to adaptive pressures—far more rapidly than other parts of the genome or the corresponding chromosomes in gorillas, for example.

In the reproductive environment we envision, characterized by frequent sexual interaction, females would typically have mated with multiple males during each ovulatory period, as do female chimps and bonobos. Thus, men with impaired fertility would have been unlikely to father children, as their sperm cells would have been overwhelmed by those of other sexual partners. The genes for robust sperm production are strongly favored in such an environment, while mutations resulting in decreased male fertility would have been filtered out of the gene pool, as they still are in chimps and bonobos.

But now consider the repercussions of culturally imposed sexual monogamy—even if enforced only for women, as was often the case until recently. In a monogamous mating system where a woman has sex with only one man, there is no sperm competition with other males. Sex becomes like an election in a dictatorship: just one candidate can win, no matter how few votes are cast. So, even a man with impaired sperm production is likely to ring the bell *eventually*, thus conceiving sons (and perhaps daughters) with increased potential for weakened fertility. In this scenario, genes associated with reduced fertility would no longer be removed from the gene pool. They'd spread, resulting in a steady reduction in overall male fertility and generalized atrophy of human sperm-production tissue.

Just as eyeglasses have allowed survival and reproduction for people with visual incapacities that would have doomed them (and their genes) in ancestral environments, sexual monogamy permits fertility-reducing mutations to proliferate, causing testicular diminishments that would never have lasted among our nonmonogamous ancestors. The most recent estimates show that sperm dysfunction affects about one in twenty men around the world, being the single most common cause of subfertilility in couples (defined as no pregnancy after a year of trying). Every indication is that the problem is growing steadily worse.[12] Nobody's maintaining the spare fridge much anymore, so it's breaking down.

If our paradigm of prehistoric human sexuality is correct, in addition to environmental toxins and food additives, sexual monogamy may be a significant factor in the contemporary infertility crisis. Widespread monogamy may also help explain why, despite our promiscuous past, the testicles of contemporary *Homo sapiens* are smaller than those of chimps and bonobos and, as indicated by our excess sperm-production capacity, than those of our own ancestors.

Sexual monogamy itself may be shrinking men's balls.

Perhaps we can declare an end to the standoff between those who argue that small human testicles tell "a story of romance and bonding between the sexes going back a long time, perhaps to the beginning of our lineage," and those who contend that our slightly-larger-than-they-should-be-if-we're-really-monogamous testicles indicate many millennia of "mild polygyny." Humans have medium-size testicles by primate standards—with strong indications of recent dwindling—but can still produce ejaculates teeming with hundreds of millions of spermatozoa. Along with a penis adapted to sperm competition, human testicles strongly suggest ancestral females had multiple lovers within a menstrual cycle. Human testicles are the equivalent of drying apples on a November tree—shrinking reminders of days gone by.

As a way of testing this hypothesis, we should find that relative penis and testicle data differ among racial and cultural groups. These differences—theoretically due to significant and consistent differences in the intensity of sperm competition in recent historical times—are what we do find, if we dare to look.[13]

Because fit is so important in the effectiveness of condoms, World Health Organization guidelines specify different sizes for various parts of the world: a 49-millimeter-width condom for Asia, a 52-millimeter width for North America and Europe, and a 53-millimeter width for Africa (all condoms are longer than most men will ever need). The condoms manufactured in China for their domestic market are 49 millimeters wide. According to a study conducted by the Indian Council of Medical Research, high levels of slippage and failure are due to a bad fit between many Indian men and the international standards used in condom manufacture.[14]

According to an article published in *Nature*, Japanese and Chinese men's testicles tend to be smaller than those of Caucasian men, on average. The authors of the study concluded that "differences in body size make only a slight contribution to these values."[15] Other researchers have confirmed these general trends, finding average combined testes weights of 24 grams for Asians, 29 to 33 grams for Caucasians, and 50 grams for Africans.[16] Researchers found "marked differences in testis size among human races. Even controlling for age differences among samples, adult male Danes have testes that are more than twice the size of their Chinese equivalents, for example."[17] This range is far beyond what average racial differences in body size would predict. Various estimates conclude that Caucasians produce about twice the number of spermatozoa per day than do Chinese ($185–235 \times 10^6$ compared with 84×10^6).

These are dangerous waters we're swimming in, dear reader, suggesting that culture, environment, and behavior can be reflected in anatomy—genital anatomy, at that. But any serious biologist or physician knows there are anatomical differences expressed racially. Despite the hair-trigger sensitivity of these issues, *not* to consider racial background in the diagnosis and treatment of disease would be unethical.

Still, some of the reluctance to connect culturally sanctioned behavior with genital anatomy is as much due to the difficulty of finding reliable historical information about true rates of female promiscuity as to the emotionally charged nature of the material itself. Furthermore, diet and environmental factors would have to be factored in before arriving at any solid conclusions regarding the relationship between sexual monogamy and genital anatomy. For example, many Asian diets include large quantities of soy products, while many Western people consume large quantities of beef, both having been shown to cause rapid generational reduction in testicular volume and spermatogenesis. Given the controversial nature of such research, and the complexities of eliminating so many variables, perhaps it's not surprising that this is an area few researchers are eager to enter.

There is a wide world of evidence that human sexual activity goes far beyond what's needed for reproduction. While the social function of sex is now seen mainly in terms of maintaining the nuclear family, this is far from the only way societies channel human sexual energies to promote social stability.

Coming in at hundreds or thousands of copulations per child born, human beings outcopulate even chimpanzees and bonobos—and are far beyond gorillas and gibbons. When the average duration of each copulation is factored in, the sheer amount of time spent in sexual activity by human beings easily surpasses that of any other primate—even if we agree to ignore all our fantasizing, dreaming, and masturbating.

The evidence that sperm competition played a role in human evolution is simply overwhelming. In the words of one researcher, "Without sperm warfare during human evolution, men would have tiny genitals and produce few sperm. . . . There would be no thrusting during intercourse, no sex dreams or fantasies, no masturbation, and we would each feel like having intercourse only a dozen or so times in our entire lives. . . . Sex and society, art and literature—in fact, the whole of human culture—would be different."[18] We can add to this list the fact that men and women would be the same height and weight (if monogamous) or that men would likely be twice the size of women (if polygynous).

Just as Darwin's famous finches in the Galápagos evolved different beak structures for cracking different seeds, related species often evolve different mechanisms for sperm competition. The sexual evolution of chimps and bonobos followed a strategy dependent upon repeated ejaculations of small but highly concentrated deposits of sperm cells, while humans evolved an approach featuring:

- a penis designed to pull back preexisting sperm, with extended, repeated thrusting;
- less frequent (compared to chimps and bonobos) but larger ejaculates;
- testicular volume and libido far beyond what's needed for monogamous or polygynous mating;
- rapid-reaction DNA controlling development of testicular tissue, this DNA apparently being absent in monogamous or polygynous primates;
- overall sperm content per ejaculate—even today—in the range of chimps and bonobos; and
- the precarious location of the testicles in a vulnerable external scrotum, associated with promiscuous mating.

In Spanish, the word *esperar* can mean "to expect" or "to hope"—depending on context. "Archaeology," writes Bogucki, "is very much constrained by what the modern imagination allows in the range of human behavior."[19] So is evolutionary theory. Perhaps so many still conclude that sexual monogamy is characteristic of our species' evolutionary past, despite the clear messages inscribed in every man's body and appetites, because this is what they *expect* and *hope* to find there.

CHAPTER EIGHTEEN

The Prehistory of O

Now there you have a sample of man's "reasoning powers," as he calls them. He observes certain facts. For instance, that in all his life he never sees the day that he can satisfy one woman; also, that no woman ever sees the day that she can't overwork, and defeat, and put out of commission any ten masculine plants that can be put to bed to her. He puts those strikingly suggestive and luminous facts together, and from them draws this astonishing conclusion: The Creator intended the woman to be restricted to one man.

Mark Twain, *Letters from the Earth*

We recently spotted a young man strolling down Las Ramblas in Barcelona proudly sporting a T-shirt proclaiming that he was *Born to F*ck.* One wonders whether he has a whole set of these shirts at home: *Born to Bre*the, Born to E*t, Born to Dr*nk, Born to Sh*t,* and of course the depressing but inevitable *Born to D*e.*

But maybe he was making a deeper point. After all, the argument central to this book is that sex has long served many crucial functions for *Homo sapiens*, with reproduction being only the most obvious among them. Since we human beings spend more time and energy planning, executing, and recalling our sexual exploits than any other species on Earth, maybe we all should wear such shirts.

Or maybe just the women. When it comes to sex, men may be trash-talking sprinters, but it's the women who win all the marathons. Any marriage counselor will tell you the most common sex-related complaint women make about men is that they are too quick and too direct. Meanwhile, men's most frequent sex-related gripe about women is that they take too damned long to get warmed up. After an orgasm, a woman may be anticipating a dozen more. A female body in motion tends to stay in motion. But men come and go. For them, the curtain falls quickly and the mind turns to unrelated matters.

This symmetry of dual disappointment illustrates the almost comical incompatibility between men's and women's sexual response in the context of monogamous mating. You have to wonder: if men and women evolved together in sexually monogamous couples for millions of years, how did we end up being so incompatible? It's as if we've been sitting down to dinner together, millennium after millennium, but half of us can't help wolfing everything down in a few frantic, sloppy minutes, while the other half are still setting the table and lighting candles.

Yes, we know: *mixed* strategies, lots of cheap sperm versus a few expensive eggs in one basket, and so on. But these flagrantly maladjusted sexual responses make far more sense when viewed as relics of our having evolved in promiscuous groups. Rather than spinning theories within theories in an effort to prop up an unstable paradigm—monogamy with *mistakes*, *mild* polygyny, *mixed* mating strategies, *serial* monogamy—can we simply face the one scenario where none of this self-contradicting, inconsistent special pleading is necessary?

Okay, fine, it's embarrassing. Maybe even humiliating, if you're prone to that sort of thing. But 150 years after *On the Origin of Species* was published, isn't it time to accept that our ancestors evolved along a sexual trajectory similar to that of our two highly social, very intelligent, closely related primate cousins? With any other question we have about the origins of human behavior, we look to chimps and bonobos for important clues: language, tool use, political alliances, war, reconciliation, altruism . . . but when it comes to sex, we prudishly turn away from these models to the distantly related, antisocial, low-I.Q. *but monogamous* gibbon? Really?

We've pointed out how the agricultural revolution triggered radical social reconfigurations from which we're still reeling. Perhaps the far-fetched denial of our promiscuous sexual prehistory expresses a legitimate fear of social instability, but insistent demands for a stable social order (based, as we're often reminded, upon the nuclear family unit) cannot erase the effects of the hundreds of thousands of years that came before our species settled into stable villages.

If female chimps and bonobos could talk, do we really think they'd be griping to their hairy girlfriends about prematurely ejaculating males who don't bring flowers anymore? Probably not, because as we've seen, when a female chimp or bonobo is in the mood, she's likely to be the center of plenty of eager male attention. And the more attention she gets, the more she attracts, because as it turns out, our male primate cousins get turned on by the sight and sound of others of their species having sex. Imagine that.

"What Horrid Extravagancies of Minde!"

> *No man (who is but never so little versed in such matters) is ignorant, what grievous symptomes, the Rising, Bearing down and Perversion, and Convulsion of the Wombe do excite; what horrid extravagancies of minde, what Phrensies, Melancholy Distempers, and Outragiousness, the preternatural Diseases of the Womb do induce, as if affected Persons were inchanted. . . .*
>
> WILLIAM HARVEY, *Anatomical Exercitations concerning the Generation of Living Creatures* (1653)

Hysteria was one of the first diseases to be described formally. Hippocrates discussed it in the fourth century BCE, and you'll find it in any medical text covering women's health written from medieval times until it was removed from the list of recognized medical diagnoses in 1952 (twenty-one years before homosexuality was finally removed). Hysteria was still one of the most diagnosed diseases in the United States and Great Britain as recently as the early twentieth century. You might wonder how physicians treated this chronic condition over the centuries.

According to historian Rachel Maines, female patients were routinely massaged to orgasm from the time of Hippocrates until the 1920s. Have a seat; the doctor will be right with you. . . .

While some passed the job off to nurses, Maines found that most physicians performed the therapy themselves, though apparently not without some difficulty. Nathaniel Highmore, writing in 1660, noted that it was not an easy technique to learn, being "not unlike that game of boys in which they try to rub their stomachs with one hand and pat their heads with the other."

Whatever challenges male physicians faced in mastering the technique, it seems to have been worth the effort. *The Health and Diseases of Women*, published in 1873, estimates that about 75 percent of American women were in need of these treatments and that they constituted the *single largest market for therapeutic services.* Despite Donald Symons's protestations that "[a]mong all peoples sexual intercourse is understood to be a service or favor that females render to males," it seems that for centuries, orgasmic release was a service male doctors rendered to women . . . for a price.

Much of this information comes from *The Technology of Orgasm*, Maines's book on this "disease" and its treatment through the centuries.[1] And what were the symptoms of this "disease"? Unsurprisingly, they were identical to those of sexual frustration and chronic arousal: "anxiety, sleeplessness, irritability, nervousness, erotic fantasy, sensations of heaviness in the abdomen, lower pelvic edema, and vaginal lubrication."

This supposed *medical* treatment for horny, frustrated women was not an isolated aberration confined to ancient history, but just one element in an ancient crusade to pathologize the demands of the female libido—a libido that experts have long insisted hardly exists.

The men who provided this lucrative therapy didn't write about "orgasm" in the medical articles they published on hysteria and its treatment. Rather, they published serious, sober discussion of "vulvular massage" leading to "nervous paroxysm" that brought temporary relief to the patient. These were ideal patients, after all. They didn't die or recover from their condition. They just kept returning, eager for more treatment sessions.

This arrangement might strike some readers as the very definition of "good work if you can get it," but many physicians apparently felt otherwise. Maines found "no evidence that male physicians enjoyed providing pelvic massage treatments. On the contrary, this male elite sought every opportunity to substitute other devices for their fingers."

What "other devices" does Maines have in mind? See if you can finish this series:

1. Sewing machine
2. Fan
3. Tea kettle
4. Toaster
5. ?

Here's a hint: These are the first five electrical appliances sold directly to American consumers. Give up? The Hamilton Beach Company of Racine, Wisconsin, patented the first home-use vibrator in 1902, thereby making it just the fifth electrical appliance approved for domestic use. By 1917, there were more vibrators than toasters in American homes. But before it became an instrument for self-treatment ("All the pleasures of youth . . . will throb within you," one suggestive ad promised), vibrators had already been in use for decades in the offices of physicians who'd grown weary of "rubbing their stomachs and patting their heads at the same time."

Motivated by the wonders of industrialization, many doctors had sought a way to mechanize the delivery of their treatment. American ingenuity would mass-produce orgasms for women who were denied them in their "properly chaste," sexually deprived lives: the first vibrators were invented by these enterprising physicians.

Late nineteenth- and early twentieth-century medical tinkerers designed all sorts of devices to provoke the necessary *nervous paroxysms* in their patients. Some were diesel powered; others ran on steam, like little locomotives that could. Some were huge contraptions hanging from the rafters on chains and pulleys, like engine blocks at an auto shop. Others sported pistons thrusting dildos through holes in tables or involved high-pressure water directed at the patient's genitalia like a fire brigade called in to douse the consuming flames of female passion. And all the while, the good doctors never publicly admitted that what they were doing was more sex than medicine.

But perhaps even more dumbfounding than their silence on being paid to provoke *nervous paroxysms* like so many Chippendale's studs is the fact that these same medical authorities managed to maintain the conviction that female sexuality was a weak and reluctant thing.

The medical monopoly on providing socially acceptable extramarital orgasms to women was assured by a strict prohibition against women or girls masturbating *themselves* to orgasm. In 1850, the *New Orleans Medical & Surgical Journal* declared masturbation public enemy number one, warning: "Neither plague, nor war, nor smallpox, nor a crowd of similar evils, have resulted more disastrously for humanity than the habit of masturbation: it is the destroying element of civilized society." Children and adults were warned that masturbation was not only sinful, but *very* dangerous—sure to result in severe health consequences, including blindness, infertility, and insanity. Besides, these authorities intoned, "normal" women had little sexual desire anyway.

In his *Psychopathia Sexualis*, published in 1886, German neurologist

Richard von Krafft-Ebing declared what everyone already thought they knew: "If [a woman] is normally developed mentally and well-bred, her sexual desire is small. If this were not so, the whole world would become a brothel and marriage and a family impossible."[2] To have suggested that women enjoyed, indeed *needed* regular orgasmic release, would have been shocking to men and humiliating to most women. Perhaps it still is.

While the anti-masturbation frenzy has roots deep in Judeo-Christian history, it found unfortunate medical support in Simon André Tissot's *A Treatise on the Disease Produced by Onanism*, published in 1758. Tissot apparently recognized the symptoms of syphilis and gonorrhea, which were considered a single disease at the time. But he misunderstood these symptoms as signs of semen depletion due to promiscuity, prostitution, and masturbation.[3]

A century later, in 1858, a British gynecologist named Isaac Baker Brown (president of the Medical Society of London at the time) proposed that most women's diseases were attributable to overexcitement of the nervous system, with the pudic nerve, which runs to the clitoris, being particularly culpable. He listed the eight stages of progressive disease triggered by female masturbation:

1. Hysteria
2. Spinal irritation
3. Hysterical epilepsy
4. Cataleptic fits
5. Epileptic fits
6. Idiocy
7. Mania
8. Death

Baker Brown argued that surgical removal of the clitoris was the best way to prevent this fatal slide from pleasure to idiocy to death. After gaining considerable celebrity and performing an unknown number of clitorectomies, Baker Brown's methods fell out of favor and he was expelled from the London Obstetrical Society in disgrace. Baker Brown subsequently went insane, and clitorectomy was discredited in British medical circles.[4]

Unfortunately, Baker Brown's writing had already had a significant impact on medical practice across the Atlantic. Clitorectomies continued to be performed in the United States well into the twentieth century as a cure for hysteria, nymphomania, and female masturbation. As late as 1936, *Holt's Diseases of Infancy and Childhood*, a respected medical-school text, recommended surgical removal or cauterization of the clitoris as a cure for masturbation in girls.

By the middle of the twentieth century, as the procedure was finally falling into disrepute in the United States, it was revived with a new rationale. Now, rather than a way to stamp out masturbation, surgical removal of large clitorises was recommended for cosmetic purposes.[5]

Before becoming a target for surgery, the clitoris had been ignored by male authors of elaborate anatomical sketchbooks for centuries. It wasn't until the mid-1500s that a Venetian professor by the name of Matteo Realdo Colombo, who had previously studied anatomy with Michelangelo, stumbled upon a mysterious *protuberance* between a woman's legs. As described in Federico Andahazi's historical novel *The Anatomist*, Colombo made this discovery while examining a patient named Inés de Torremolinos. Colombo noted that Inés grew tense when he manipulated this small button, and that it appeared to grow in size at his touch. Clearly, this would require further exploration. After examining scores of other women, Colombo found that all of them had this same heretofore "undiscovered" protuberance and that they all responded similarly to gentle manipulation.

In March of 1558, Andahazi tells us that Colombo proudly reported his "discovery" of the clitoris to the dean of his faculty.[6] As Jonathan Margolis speculates in *O: The Intimate History of the Orgasm*, the response was probably not what Colombo had anticipated. The professor was "arrested in his classroom within days, accused of heresy, blasphemy, witchcraft and Satanism, put on trial and imprisoned. His manuscripts were confiscated, and his [discovery] was never permitted to be mentioned again until centuries after his death."[7]

Beware the Devil's Teat

The "illness" that led frustrated women to the offices of vibrator-wielding doctors a century ago often led someplace far worse in medieval Europe. As historian Reay Tannahill explains, "The *Malleus Maleficarum* (1486), the first great handbook of the witch inquisitors, had no more difficulty than a modern psycho-analyst in accepting that [a certain] type of woman might readily believe she had had intercourse with the Devil himself, a huge, black, monstrous being with an enormous penis and seminal fluid as cold as ice water."[8] But it wasn't only sexual dreams that attracted the brutal attentions of erotophobic authorities. If a witch-hunter in the 1600s discovered a woman or girl with an unusually large clitoris, this "devil's teat" was sufficient to condemn her to death.[9]

Medieval Europe suffered periodic plagues of incubi and succubi, male and female demons thought to be invading the dreams, beds, and bodies of living people. Thomas Aquinas and others believed that these demons impregnated women on their nocturnal visits by first posing as a succubus (a female spirit who has sex with a sleeping man in order to obtain his sperm), and then depositing the sperm in an unsuspecting woman in the form of an incubus (a male spirit ravishing a sleeping woman). Women thus thought to have been impregnated by malevolent spirits flitting about like nocturnal honeybees were at special risk of being exposed as witches and dealt with accordingly. Any stories these women might have told regarding the true origins of their pregnancy conveniently died with them.

Though now considered one of the finest novels ever written, *Madame Bovary* was denounced as immoral when it was first published in late 1856. Public prosecutors in Paris were upset that Gustave Flaubert portrayed a headstrong peasant girl who flaunted the rules of established propriety by taking lovers. They felt her character met with insufficient punishment. Flaubert's defense was that the work was "eminently moral" on those terms. After all, Emma Bovary dies by her own hand in misery, poverty, shame, and desperation. Insufficient punishment? The case against the book, in other words, turned on whether Emma Bovary's punishment was agonizing and horrible *enough*, not on whether she deserved such suffering at all or had any right to pursue sexual fulfillment in the first place.

But even Flaubert and his misogynistic prosecutors could never have dreamed up the punishments said to befall immodest women among the Tzotzil Maya of Central America. Sarah Blaffer Hrdy explains that "the h'ik'al, a super-sexed demon with a several-foot-long penis," seizes women who have misbehaved, "carrying them off to his cave, where he rapes them." Little girls are told that any woman unlucky enough to become pregnant by the h'ik'al "swells up and then gives birth night after night, until she dies."[10]

This apparent need to punish female sexual desire as something evil, dangerous, and pathological is not limited to medieval times or remote Mayan villages. Recent estimates by the World Health Organization suggest that more than 100 million girls and women are living with the consequences of genital mutilation.

The Force Required to Suppress It

> *A fire is never sated by any amount of logs, nor the ocean by rivers that flow into it; death cannot be sated by all the creatures in the world, nor a fair-eyed woman by any amount of men.*
>
> THE KAMA SUTRA

Before the war on drugs, the war on terror, or the war on cancer, there was the war on female sexual desire. It's a war that has been raging far longer than any other, and its victims number well into the billions by now. Like the others, it's a war that can never be won, as the declared enemy is a force of nature. We may as well declare war on the cycles of the moon.

There is a pathetic futility animating the centuries-long insistence—against overwhelming evidence to the contrary—that the human female is indifferent to the insistent urgings of libido. Recall the medical authorities in the antebellum South who assured plantation owners that slaves trying to break out of their chains were not human beings deserving of freedom and dignity, but sufferers of *Drapetomania*, a medical disorder best cured with a good lashing. And who can forget the "well-intentioned" Inquisition that forced Galileo to disown truths as obvious to him as they were offensive to minds calcified by power and doctrine? In this ongoing struggle between what *is* and what many post-agricultural patriarchal societies insist *must be*, women who have dared to renounce the credo of the coy female are still spat upon, insulted, divorced, separated from their children, banished, burned as witches, pathologized as hysterics, buried to their necks in desert sand, and stoned to death. They and their children—those "sons and daughters of bitches"—are still sacrificed to the perverse, conflicted gods of ignorance, shame, and fear.

If psychiatrist Mary Jane Sherfey was correct when she wrote, "The strength of the drive determines the force required to suppress it" (an observation downright Newtonian in its irrefutable simplicity), then what are we to make of the force brought to bear on the suppression of female libido?[11]

CHAPTER NINETEEN

When Girls Go Wild

Female Copulatory Vocalization

Here's a question we ask the audience every time we give a public presentation: If you've ever heard a heterosexual couple having sex (and who hasn't?), which partner was louder? The answer we get every time, every place—from men, women, straight, gay, American, French, Japanese, and Brazilians—is always the same. Hands down. No question about it. Not even close. We don't have to tell you because you already know, don't you? Yes, the "meek," "demure," "coy" sex is the source of the high-decibel moaning, groaning, and calling out to the good Lord above, neighbors be damned.

But why? Within the framework of the standard narrative of human sexuality, what scientists call *female copulatory vocalization* (FCV) is a major conundrum. You'll recall Steven Pinker's claim that "[i]n all societies, sex is at least somewhat 'dirty.' It is conducted in private. . . ."[1] Why would the female of such a species risk attracting all that attention? Why is it that from the Lower East Side to the upper reaches of the Amazon, women are far more likely than men to loudly announce their sexual pleasure for all to hear?

And why is the sound of a woman having an orgasm so difficult for heterosexual men to ignore?[2] They say women can hear a baby crying from a great distance, but gentlemen, we ask you, is there any sound easier to pick out of the cacophony of an apartment block—and harder to ignore—than that of a woman lost in passion?

If you're one of the ten or fifteen people alive who have never seen Meg Ryan's fake orgasm scene in *When Harry Met Sally*, go watch it now (it's easily accessible online). It's one of the best-known scenes in all of modern cinema, but if the roles were reversed, the scene wouldn't be funny—it wouldn't even make sense. Imagine: Billy Crystal sits at the restaurant table, he starts breathing harder, maybe his eyes bug out a bit, he grunts a few times, takes a few bites of his sandwich, and falls asleep. No big laughs. Nobody in the deli even notices. If male orgasm is a muffled crash of cymbals, female orgasm is full-on opera. Full of screaming, shouting, singing people standing around with spears, and table pounding sure to quiet even the noisiest New York deli.[3]

Female cries of ecstasy aren't a modern phenomenon. The *Kama Sutra* contains ancient advice on female copulatory vocalization in terms of erotic technique, categorizing an aviary of ecstatic expression a woman might choose from: "As a major part of moaning, she may use, according to her imagination, the cries of the dove, cuckoo, green pigeon, parrot, bee, nightingale, goose, duck and partridge." *A goose?* Honk if You ♥ Sex!

But apart from barnyard erotic technique, it just doesn't make sense for the female of a monogamous (or "mildly polygynous") species to call attention to herself when mating. On the other hand, if thousands of generations of multiple mating are built into modern human sexuality, it's pretty clear what all the shouting's about.

As it turns out, women aren't the only female primates making a lot of noise in the throes of passion. British primatologist Stuart Semple found that, "In a wide variety of species, females vocalize just before, during or immediately after they mate. These vocalizations," Semple says, "are particularly common among the primates and evidence is now accumulating that by calling, a female incites males in her group. . . ."[4]

Precisely. There's a good reason the sound of a woman enjoying a sexual encounter entices a heterosexual man. Her "copulation call" is a potential invitation to *come hither*, thus provoking sperm competition.

Semple recorded more than 550 copulation calls from seven different female baboons and analyzed their acoustic structure. He found that these complex vocalizations contained information related both to the female's reproductive state (the vocalizations were more complex when females were closer to ovulation) and to the status of the male "inspiring" any given vocalization (calls were longer and contained more distinct sonic units during matings with higher-ranked males). Thus, in these baboons at least, listening males could presumably gain information as to their likelihood of impregnating a calling female, as well as some sense of the rank of the male they'd find with her if they approached.

Meredith Small agrees that the copulation calls of female primates are easily identifiable: "Even the uninitiated can identify female nonhuman primate orgasm, or sexual pleasure. Females," Small tells us, "make noises not heard in any other context but mating."[5] Female lion-tailed macaques use copulation calls to invite male attention even when not ovulating. Small reports that among these primates, ovulating females most often directed their invitations at males outside their own troop, thus bringing new blood into the mating mix.[6]

Female copulatory vocalization is highly associated with promiscuous mating, but not with monogamy. Alan Dixson has noted that the females of promiscuous primate species emit more complex mating calls than females of monogamous and polygynous species.[7] Complexity aside, Gauri Pradhan and his colleagues conducted a survey of copulation calls in a variety of primates and found that "variation in females' promiscuity predicts their tendency to use copulation calls in conjunction with mating." Their data show that higher levels of promiscuity predict more frequent copulation calls.[8]

William J. Hamilton and Patricia C. Arrowood analyzed the copulatory vocalizations of various primates, including three human couples going at it.[9] They noticed that "female sounds gradually intensified as orgasm approached and at orgasm assumed a rapid, regular (equal note lengths and inter-note intervals) rhythm absent in the males' calls at orgasm." Still, the authors can't help sounding a tad let down when they note, "Neither sex [of human] . . . showed the complexity of note structure characteristic of baboon copulatory vocalizations." But that's probably a good thing, because elsewhere in their article we learn that female baboons' copulation calls are clearly audible to even human ears from three hundred meters away.

Before you conclude that female copulatory vocalization is just a fancy phrase for a little excitement, think about the predators possibly alerted by this primate passion. Chimps and bonobos may be out of reach up in the branches, but baboons (like our ground-dwelling ancestors) live among leopards and other predators who would be quite interested in a two-for-one special on fresh primate—especially given a mating pair's distracted, vulnerable state.

As Hamilton and Arrowood put it: "In spite of the risk of exposure of individuals and the troop to predators these baboons habitually call during copulation, [so] the calls must have some adaptive value." What could that be? The authors offered several hypotheses, including the notion that the calls may be a stratagem to help activate the male's ejaculatory reflex, an analysis with which many prostitutes would presumably agree. Perhaps there is something to this idea,[10] but even so, male primates are not known for needing a great deal of assistance in *activating their ejaculatory reflex*. If anything, the human male ejaculatory reflex tends to be too easily activated—at least from the perspective of women *not* being paid to activate it as quickly as possible. Especially given all the other convergent evidence, it seems far more likely that in humans, female copulatory vocalization would serve to attract males to the ovulating, sexually receptive female, thus promoting sperm competition, with all its attendant benefits—both reproductive and social.

Yet despite all the loud carrying on by women the world over, "The credo of the coy female persists," writes Natalie Angier. "It is garlanded with qualifications and is admitted to be an imperfect portrayal of female mating strategies but then, that little matter of etiquette attended to, the credo is stated once again."

Sin Tetas, No Hay Paraíso[11]

For better or worse, the human female's naughty bits don't swell up to five times their normal size and turn bright red to signal her sexual availability. But is there anatomical evidence suggesting that women evolved to be highly sexual? No question. It turns out that every bit as much as a man's body, the woman's body (and preconscious behavior) is replete with indications of millennia of promiscuity and sperm competition.

Considering its almost total lack of muscle tissue, the female breast wields amazing power. Curvaceous women have leveraged this power to manipulate even the most accomplished, disciplined men for as long as anyone's been around to notice. Empires have fallen, wills have been revised, millions of magazines and calendars sold, Super Bowl audiences scandalized . . . all in response to the mysterious force emanating from what are, after all, small bags of fat.

One of the oldest human images known, the so-called Venus of Willendorf, created about 25,000 years ago, features a bosom of Dolly Parton-esque dimensions. Two hundred fifty centuries later, the power of the exaggerated breast shows little sign of getting old. According to the American Society of Plastic Surgery, 347,254 breast augmentation procedures were performed in the United States in 2007, making it the nation's most commonly performed surgical procedure. What gives the female breast such transcendent influence over heterosexual male consciousness?

First, let's dispense with any purely utilitarian interpretations. While the mammary glands contained in women's breasts exist for the feeding of infants, the fatty tissue that confers the magical curve of the human breast—the swell, sway, and jiggle—has nothing to do with milk production. Given the clear physiological costs of having pendulous breasts (back strain, loss of balance, difficulty running), if they aren't meant to advertise milk for babies, why did human females evolve and retain these cumbersome appendages?

Theories range from the belief that breasts serve as signaling devices announcing fertility and fat deposits sufficient to withstand the rigors of pregnancy and breastfeeding[12] to "genital echo theory": females developed pendulous breasts around the time hominids began walking upright in order to provoke the excitation males formerly felt when gazing at the fatty deposits on the buttocks.[13] Theorists supporting genital echo theory have noted that swellings like those of chimpanzees and bonobos would interfere with locomotion in a bipedal primate, so when our distant ancestors began walking upright, they reason that some of the female's fertility signaling moved from the rear office, as it were, to the front showroom. In a bit of historical ping-pong, the dictates of fashion have moved the swelling back and forth over the centuries with high heels, Victorian bustles, and other derrière enhancements.

The visual similarity between these two bits of female anatomy has been facilitated by the recent popularity of low-cut jeans that teasingly reveal the nether cleavage. "The butt crack is the new cleavage," writes journalist Janelle Brown, "reclaimed to peek seductively from the pants of supermodels and commoners alike. . . . It's naughty and slightly tawdry," she continues, "but with the soft round charm of a perfect pair of breasts."[14] If your moon is waning, you can always don a "butt bra" from Bubbles Bodywear, which promises to create the effect that's been turning male heads since before men existed. Like the Victorian bustle, the butt bra mimics the full curves of the ovulating chimp or bonobo. Speaking of waning moons, it's worth noting that unless her breasts are artificially enhanced, as a woman's fertility fades with age, so do her breasts—further supporting the claim that they evolved to signal fertility.

Female bonobo. Photo: www.friendsofbonobos.org

Victorian bustle. Photo: Strawbridge & Clothier's Quarterly (Winter 1885–86)

The Butt Bra. Photo: Sweet and Vicious LLC
Company slogan: "Take Your Gluteus to the Maximus!"

Human females aren't the only primates with fertility signals on their chests. The Gelada baboon is another vertically oriented primate with sexual swellings on the females' chests. As we'd expect, the Gelada's swellings come and go with the females' sexual receptivity. As the human female is potentially *always* sexually receptive, her breasts are more or less always swollen, from sexual maturity on.[15]

But not all female primates have genital swellings that visually announce their ovulatory status. Meredith Small reports that only fifty-four of the seventy-eight species surveyed "experience easily seen morphological changes during cycles," and that half of these showed "only slight pinkness." Once again, our two closest primate cousins stand out from the pack in terms of their decidedly indiscreet sexuality, being the only apes with such extravagant, brightly colored sexual swellings. The female chimp's red-light district comes and goes, reflecting the waxing and waning of her fertility, but as Small confirms, the bonobo's "swellings never change much, so that bonobo females always give a signal of fertility—much as humans do."[16]

Although many theories claim the human female has "hidden ovulation," it's not really hidden at all, if you know how and where to look. Martie Haselton and her colleagues found that men shown photographs of the same thirty women—some taken around ovulation and others not—were quite good at judging when the women were "trying to look more attractive," which in turn corresponded to the women's menstrual status. These authors found that women tend to dress more fetchingly when they are more likely to be fertile. "Moreover," writes Haselton, "the closer women were to ovulation when photographed in the fertile window, the more frequently their fertile photograph was chosen."[17]

Other researchers have found that men preferred women's bodily smells near ovulation and that women tend to behave more provocatively in various ways when they're likely to be fertile (they wear more jewelry and perfume, go out more, are more likely to hook up for casual sexual encounters, and are less likely to use condoms with new lovers).

Come Again?

Much as women's breasts have fascinated evolution-minded theorists, the female orgasm has confounded them. Like breasts, female orgasm is a major head-scratcher for mainstream narratives of human sexual evolution. It's not necessary for conception, so why should it exist at all? For a long time, scientists claimed that women were the only female animals to experience orgasm. But once female biologists and primatologists arrived on the scene, it became obvious that *many* female primates were having orgasms.

The underlying motivation for claiming that female orgasm was unique to human beings probably lay in the role it played in the standard narrative. According to this view, orgasm evolved in the human female to facilitate and sustain the long-term pair bond at the heart of the nuclear family.[18] Once you've swallowed that story, it becomes problematic to admit that the females of other primate species are orgasmic, too. Your problem gets worse if the most orgasmic species happen to be the most promiscuous as well, which appears to be the case.

As Alan Dixson writes, this monogamy-maintenance explanation for female orgasm "seems far-fetched. After all," he writes, "females of other primate species, and particularly those with multimale–multifemale [promiscuous] mating systems such as macaques and chimpanzees, exhibit orgasmic responses in the absence of such bonding or the formation of stable family units." On the other hand, Dixson goes on to note, "Gibbons, which are primarily monogamous, do not exhibit obvious signs of female orgasm."[19] Although Dixson classifies humans as mildly polygynous in his survey of primate sexuality, he seems to have doubts, as when he writes, "One might argue that . . . the female's orgasm is rewarding, increases her willingness to copulate with a variety of males rather than one partner, and thus promotes sperm competition."[20]

Donald Symons and others have argued that "orgasm is most parsimoniously interpreted as a potential all female mammals possess." What helps realize this "potential" in some human societies, argues Symons, are "techniques of foreplay and intercourse [that] provide sufficiently intense and uninterrupted stimulation for females to orgasm."[21] In other words, Symons thinks women have more orgasms than mares simply because men are better lovers than stallions. Stomp your foot three times if you believe this.

In support of his theory, Symons cites studies like Kinsey's showing that fewer than half of women questioned (Americans in the 1950s) experienced orgasm at least nine out of ten times they had intercourse, whereas in other societies (he refers to Mangaia, in the South Pacific), elaborate and extended sexual play result in nearly universal orgasm for women. "Orgasm," Symons concludes, "never is considered to be a spontaneous and inevitable occurrence for females as it always is for males." For Symons, Stephen Jay Gould, Elisabeth Lloyd,[22] and others, *some* women have orgasms *sometimes* because *all* men do *every time*. For them, the female orgasm is the equivalent of male nipples: a structural echo without function in one sex of a trait vital in the other.

Given all the energy required to get there, it's surprising that the female reproductive tract is not a particularly welcoming place for sperm cells. Researchers Robin Baker and Mark Bellis found that approximately 35 percent of the sperm are ejected within half an hour of intercourse and those that remain are anything but home free.[23] The female's body perceives sperm as antigens (foreign bodies) that are promptly attacked by anti-sperm leucocytes, which outnumber sperm 100:1. Only one in 14 million ejaculated human sperm even reach the oviduct.[24] In addition to the obstacles imposed by the female's body, even those lucky few sperm are going to run into competition from other males (at least, if our model of human sexuality has any validity).

But while presenting obstacles to most sperm, the woman's body can assist others. There is striking evidence that the female reproductive system is capable of making subtle judgments based upon the chemical signature of different men's sperm cells. These assessments may go well beyond general health to the subtleties of immunological compatibility. The genetic compatibility of different men with a given woman means that sperm quality is a *relative characteristic*. Thus, as Anne Pusey explains, "Females may benefit from sampling many males, and different females will not necessarily benefit from mating with the same 'high quality' male."[25]

This is a crucially important point. Not every "high quality" male would be a good match for any specific woman—even on a purely biological level. Because of the complexities of how the two sets of parental DNA interact in fertilization, a man who *appears* to be of superior mate value (square jaw, symmetrical body, good job, firm handshake, Platinum AMEX card) may in fact be a poor genetic match for a particular woman. So, a woman (and ultimately, her child) may benefit by "sampling many males" and letting her body decide whose sperm fertilizes her. Her body, in other words, might be better informed than her conscious mind.

So, in terms of reproduction, the "fitness" of our prehistoric male ancestors was not decided in the external social world, where conventional theories tell us men competed for mates in struggles for status and material wealth. Rather, *paternity was determined in the inner world of the female reproductive tract where every woman is equipped with mechanisms for choosing among potential fathers at a cellular level.* Remember this next time you read something like, "The predisposition for influence, substance and prestige are all merely expressions of a male positioning himself to acquire women with whom to mate," or, "Mate competition will involve contests over resources [men's] wives will need to raise children."[26] This may well be the situation for most people today, but our bodies suggest our ancestors faced an entirely different scenario.

Sperm competition is best understood not as a sprint to the egg, but a race over hurdles. Aside from the anti-sperm leucocytes mentioned previously, anatomical and physiological obstacles are in the vagina, cervix, and on the surface of the ovum itself. The complexity of the human cervix suggests it evolved to filter the sperm of various males. Concerning macaques (highly promiscuous monkeys) and humans, Dixson writes: "In the genus *Macaca*, all species of which are considered to have multimale-multifemale mating systems, the cervix is especially complex in structure. . . . The evidence pertaining to human beings and macaque females," he continues, "indicates that the cervix acts both as a filtering mechanism and as a temporary reservoir for spermatozoa during their migration into the uterus."[27] As with the complex penis and external testicles in the male, the elaborate filtering design of the human cervix points toward promiscuity in our ancestors.

The idea that female choice (conscious or not) can happen *after* or *during* intercourse rather than as part of an elaborate precopulatory courtship ritual turns the standard narrative inside out and upside down. If the female's reproductive system has evolved intricate mechanisms for filtering and rejecting the sperm cells of some men while helping along those of a man who meets criteria of which she may be utterly unaware, Darwin's "coy female" starts looking like what she is: an anachronistic male fantasy.

But Darwin may have suspected more than he let on concerning postcopulatory mechanisms of sexual selection. Any discussion of human sexual behavior or the evolutionary implications of our genital morphology would have been extremely controversial in 1871, to put it mildly. Just imagine, as Dixson does, "what would have occurred if *The Descent of Man* had included a detailed exposition of the evolution of the penis and testes or descriptions of the various copulatory postures and patterns employed by animals and human beings."[28]

No one can blame Darwin for opting not to include chapters on the evolution of the penis and vagina in his already explosive work. But a century and a half is a long time for discretion and cultural bias to keep smothering scientific fact. To Meredith Small, the story of the female's role in conception is a miniature of the overall narrative. She sees the popular understanding of conception as "an outdated allegory of human sexuality" featuring the male as "aggressor, persuader, conqueror." Recent research on human fertilization suggests something of a role reversal. Small suggests the ovum "reaches out and envelops reluctant sperm." "Female biology," she concludes, "even at the level of egg and sperm interaction, doesn't necessarily dictate a docile stance."[29]

In addition to enveloping eggs, a cervix that filters or favors sperm, and vaginal contractions that may expel the sperm of one man while boosting that of another, women's orgasms provoke changes in vaginal acidity. These changes appear to assist the sperm cells of the lucky guy who provoked the orgasm. The environment at the cervical opening tends to be highly acidic and thus hostile to sperm cells. The alkaline pH of semen protects the spermatozoa in this environment for a while, but the protection is short-lived; most of the sperm cells are viable within the vagina for only a few hours, so these changes in acidity alter the vaginal environment in ways that can favor sperm that arrive with the female's orgasm.

The benefits may run both ways. Recent research suggests women who do not use condoms are less likely to suffer from depression than either women who do use condoms or who are not sexually active. Psychologist Gordon Gallup's initial survey of 293 women (data congruent with those from another survey still to be published that included 700 women) found that women can develop a "chemical dependency" on the boost they get from the testosterone, estrogen, prostaglandins, and other hormones contained in semen. These chemicals enter the woman's bloodstream through the vaginal wall.[30]

If it's true that multiple mating was common in human evolution, the apparent mismatch between the relatively quick male orgasmic response and the so-called "delayed" female response makes sense (note how the female response is "delayed" only if the male's is assumed to be "right on time"). The male's quick orgasm lessens the chances of being interrupted by predators or other males (survival of the quickest!), while the female and her child would benefit by exercising some preconscious control over which spermatozoa would be most likely to fertilize her ovum.

Prolactin and the other hormones released at orgasm appear to trigger very different responses in men and women. While a man is likely to require a prolonged refractory (or recovery) period immediately after an orgasm (and maybe a sandwich and a beer as well), thus getting him out of the way of other males, many women are willing and able to continue sexual activity well beyond a "starter orgasm."

It's worth repeating that primate species with orgasmic females tend to be promiscuous. Given the great variability of mating behavior—even just among the apes—this is highly significant. While monogamous gibbons have rarely been seen copulating, so infrequent and silent is their intercourse, female chimps and bonobos go wild regularly and shamelessly. Females often mate with every male they can find, copulating far more than is necessary for reproduction. Goodall reported seeing one female at Gombe who mated fifty times in a single day.

Echoing the *Kama Sutra*, Sherfey isn't shy about the implications of this mismatch of orgasmic capacity between human males and females, writing: "The sexual hunger of the female, and her capacity for copulation completely exceeds that of any male," and, "To all intents and purposes, *the human female is sexually insatiable. . . .*" That may or may not be, but it cannot be denied that the design of the human female's reproductive system is far from what the standard narrative predicts, and thus demands radical rethinking of the evolution of female sexuality.

PART V

Men Are from Africa, Women Are from Africa

The sooner we accept the basic differences between men and women, the sooner we can stop arguing about it and start having sex!

Dr. Stephen T. Colbert, D.F.A.

Permeating the standard narrative of human sexual evolution is the depressing claim that men and women always have been and always will be locked in erotic conflict. The War Between the Sexes is said to be built into our evolved sexuality: men want lots of no-strings lovers, while women want just a few partners, with as many strings as possible. If a man agrees to be roped into a relationship, the narrative tells us, he'll be hell-bent on making sure his mate isn't risking his genetic investment by accepting *deposits* from other men, as it were.

Extreme as it sounds, this is no overstatement. In his classic

1972 paper on "parental investment," biologist Robert Trivers remarked, "One can, in effect, treat the sexes as if they were different species, the opposite sex being a resource relevant to producing maximum surviving offspring." In other words, men and women have such conflicting agendas when it comes to reproduction that we are essentially *predators* of one another's interests. In *The Moral Animal*, Robert Wright laments, "A basic underlying dynamic between men and women is mutual exploitation. They seem, at times, designed to make each other miserable."[1]

Don't believe it. We aren't designed to make each other miserable. This view holds evolution responsible for the mismatch between our evolved predispositions and the post-agricultural socioeconomic world we find ourselves in. The assertion that human beings are naturally monogamous is not just a lie; it's a lie most Western societies insist we keep telling each other.

There's no denying that men and women are different, but we're hardly different species or from different planets or *designed* to torment one another. In fact, the interlocking nature of our differences testifies to our profound mutuality. Let's look at some of the ways in which male and female erotic interests, perspectives, and capacities converge, intersect, and overlap, showing how each of us is a fragment of a greater unity.

CHAPTER TWENTY

On Mona Lisa's Mind

Do I contradict myself?
Very well then I contradict myself,
(I am large, I contain multitudes.)

Walt Whitman, *Song of Myself*

Faced with the mysteries of woman, Sigmund Freud, who seemed to have an answer for everything else, came up empty. "Despite my thirty years of research into the feminine soul," he wrote, "I have not yet been able to answer . . . the great question that has never been answered: what does a woman want?"

It's no accident that what the BBC called "the most famous image in the history of art" is a study of the inscrutable feminine created by a homosexual male artist. For centuries, men have been wondering what Leonardo da Vinci's *Mona Lisa* was thinking. Is she smiling? Is she angry? Disappointed? Unwell? Nauseated? Sad? Shy? Turned on? None of the above?

Probably closer to *all of the above.* Does she contradict herself? Very well, then. The *Mona Lisa* is large. Like all women, but more—like all that is feminine—she reflects every phase of the moon. She contains multitudes.

Our journey into a deeper understanding of the "feminine soul" begins in a muddy field in the English countryside. In the early 1990s, neuroscientist Keith Kendrick and his colleagues exchanged that season's newborn sheep and goats (the baby sheep were raised by adult goats, and vice versa). Upon reaching sexual maturity a few years later, the animals were reunited with their own species and their mating behavior was observed. The females adopted a love-the-one-you're-with approach, showing themselves willing to mate with males of either species. But the males, *even after being back with their own species for three years*, would mate *only* with the species with which they were raised.[1]

Research like this suggests strong differences in degrees of "erotic plasticity" (changeability) in the males and females of many species—including ours.[2] The human female's sexual behavior is typically far more malleable than the male's. Greater erotic plasticity leads most women to experience more variation in their sexuality than men typically do, and women's sexual behavior is far more responsive to social pressure. This greater plasticity could manifest through changes in whom a woman wants, in how much she wants him/her/them, and in how she expresses her desire. Young males pass through a brief period in which their sexuality is like hot wax waiting to be imprinted, but the wax soon cools and solidifies, leaving the imprint for life. For females, the wax appears to stay soft and malleable throughout their lives.

This greater erotic plasticity appears to manifest in women's more holistic responses to sexual imagery and thoughts. In 2006, psychologist Meredith Chivers set up an experiment where she showed a variety of sexual videos to men and women, both straight and gay. The videos included a wide range of possible erotic configurations: man/woman, man/man, woman/woman, lone man masturbating, lone woman masturbating, a muscular guy walking naked on a beach, and a fit woman working out in the nude. To top it all off, she also included a short film clip of bonobos mating.[3]

While her subjects were being buffeted by this onslaught of varied eroticism, they had a keypad where they could indicate how turned on they felt. In addition, their genitals were wired up to plethysmographs. *Isn't that illegal?* No, a plethysmograph isn't a torture device (or a dinosaur, for that matter). It measures blood flow to the genitals, a surefire indicator that the body is getting ready for love. Think of it as an erotic lie detector.

What did Chivers find? Gay or straight, the men were predictable. The things that turned them on were what you'd expect. The straight guys responded to anything involving naked women, but were left cold when only men were on display. The gay guys were similarly consistent, though at 180 degrees. And both straight and gay men indicated with the keypad what their genital blood flow was saying. As it turns out, men *can* think with both heads at once, as long as both are thinking the same thing.

The female subjects, on the other hand, were the very picture of inscrutability. Regardless of sexual orientation, most of them had the plethysmograph's needle twitching over just about everything they saw. Whether they were watching men with men, women with women, the guy on the beach, the woman in the gym, or bonobos in the zoo, their genital blood was pumping. But unlike the men, many of the women reported (via the keypad) that they weren't turned on. As Daniel Bergner reported on the study in *The New York Times*, "With the women . . . mind and genitals seemed scarcely to belong to the same person."[4] Watching both the lesbians and the gay male couple, the straight women's vaginal blood flow indicated more arousal than they confessed on the keypad. Watching good old-fashioned vanilla heterosexual couplings, everything flipped and they claimed *more* arousal than their bodies indicated. Straight or gay, the women reported almost no response to the hot bonobo-on-bonobo action, though again, their bodily reactions suggested they kinda liked it.

This disconnect between what these women experienced on a physical level and what they consciously registered is precisely what the theory of differential erotic plasticity predicts. It could well be that the price of women's greater erotic flexibility is more difficulty in knowing—and, depending on what cultural restrictions may be involved, in accepting—what they're feeling. This is worth keeping in mind when considering why so many women report lack of interest in sex or difficulties in reaching orgasm.*

If you aren't confused already, consider that research psychiatrist Andrey Anokhin and his colleagues found that erotic images elicit significantly quicker and stronger response in women's brains than either pleasant or frightening images without erotic content. They showed 264 women a randomly ordered collection of images ranging from snarling dogs to water skiers to semi-naked couples getting hot and heavy. The women's brains responded about 20 percent faster to the erotic images than to any others. With men this eager responsiveness was expected, but the results among the supposedly less visual, less libidinous women surprised the researchers.[5]

The female erotic brain is full of such surprises. Dutch researchers used positron emission tomography (PET) to scan the brains of thirteen women and eleven men in the throes of orgasm. While the brevity of the male orgasm made reliable readings difficult to get, the heightened activity they found in the secondary somatosensory cortex (associated with genital sensation) was what they'd expected. But the women's brains left the researchers befuddled. It seems the female brain goes into standby mode at orgasm. What little increase in cerebral activity the ladies' brains exhibited was in the primary somatosensory cortex, which registers the presence of sensation, but not much excitement about it. "In women the primary feeling is there," one of the researchers said, "but not the marker that this is seen as a big deal. For males, the touch itself is all-important. For females, it is not so important."[6]

Every woman knows her menstrual cycle can have profound effects on her eroticism. Spanish researchers confirmed that women experience greater feelings of attractiveness and desire around ovulation, while

* This disconnect is also relevant to the research on jealousy we discussed in Chapter 10.

others have reported that women find classically masculine faces more attractive around ovulation, opting for less chiseled-looking guys when not fertile.[7] Since the birth control pill affects the menstrual cycle, it's not surprising that it may affect a woman's patterns of attraction as well. Scottish researcher Tony Little found women's assessment of men as potential husband material shifted if they were on the pill. Little thinks the social consequences of his finding may be immense: "Where a woman chooses her partner while she is on the Pill, and then comes off it to have a child, her hormone-driven preferences have changed and she may find she is married to the wrong kind of man."[8]

Little's concern is not misplaced. In 1995, Swiss biological researcher Claus Wedekind published the results of what is now known as the "Sweaty T-shirt Experiment." He asked women to sniff T-shirts men had been wearing for a few days, with no perfumes, soaps, or showers. Wedekind found, and subsequent research has confirmed, that most of the women were attracted to the scent of men whose major histocompatibility complex (MHC) differed from her own.[9] This preference makes genetic sense in that the MHC indicates the range of immunity to various pathogens. Children born of parents with different immunities are likely to benefit from a broader, more robust immune response themselves.

The problem is that women taking birth control pills don't seem to show the same responsiveness to these male scent cues. Women who were using birth control pills chose men's T-shirts randomly or, even worse, showed a preference for men with similar immunity to their own.[10]

Consider the implications. Many couples meet when the woman is on the pill. They go out for a while, like each other a lot, and then decide to get together and have a family. She goes off the pill, gets pregnant, and has a baby. But her response to him changes. There's something about him she finds irritating—something she hadn't noticed before. Maybe she finds him sexually unattractive, and the distance between them grows. But her libido is fine. She gets flushed every time she gets close enough to smell her tennis coach. Her body, no longer silenced by the effects of the pill, may now be telling her that her husband (still the great guy she married) isn't a good genetic match for her. But it's too late. They blame it on the work pressure, the stress of parenthood, each other. . . .

Because this couple inadvertently short-circuited an important test of biological compatibility, their children may face significant health risks ranging from reduced birth weight to impaired immune function.[11] How many couples in this situation blame themselves for having "failed" somehow? How many families are fractured by this common, tragic, undetected sequence of events?[12]

Psychologist Richard Lippa teamed up with the BBC to survey over 200,000 people of all ages from all over the world concerning the strength of their sex drive and how it affects their desires.[13] He found the same inversion of male and female sexuality: for men, both gay and straight, higher sex drive increases the specificity of their sexual desire. In other words, a straight guy with a higher sex drive tends to be more focused on women, while higher sex drive in a gay guy makes him more intent on men. But with women—at least nominally straight women—the *opposite* occurs: the higher her sex drive, the more likely she'll be attracted to men *and* women. Lesbians showed the same pattern as men: a higher sex drive means more women-only focus. Perhaps this explains why nearly twice as many women as men consider themselves bisexual, while only half as many consider themselves to be exclusively gay.

Those who claim this just means men are more likely to be repressing some universal human bisexuality will have to consider sexologist Michael Bailey's fMRI scans of gay and straight men's brains while they viewed pornographic photos. They reacted as men tend to do: simply and directly. The gay guys liked the photos showing men with men, while the straight guys were into the photos featuring women. Bailey was looking for activation of the brain regions associated with inhibition, to see whether his subjects were denying a bisexual tendency. No dice. Neither gay nor straight men showed unusual activation of these regions while viewing the photos. Other experiments using subliminal images have generated similar results: gay men, straight men, and lesbians all responded as predicted by their stated sexual orientation, while nominally straight women ("I contain multitudes") responded to just about everything. This is just how we're wired, not the result of repression or denial.[14]

Of course, signs of repression aren't hard to find in sex research. There's plenty. For example, one of the long-standing mysteries of human sexuality has been that heterosexual men tend to report having more sexual encounters and partners than heterosexual women do—a mathematical impossibility. Psychologists Terry Fisher and Michele Alexander decided to take a closer look at people's claims regarding age of first sexual experience, number of partners, and frequency of sexual encounters.[15] Fisher and Alexander set up three different testing conditions:

1. The subjects were led to believe their answers might be seen by the researchers waiting just outside the room.
2. The subjects could answer the questions privately and anonymously.
3. The subjects had electrodes placed on their hand, arm, and neck—believing themselves (falsely) to be hooked up to a lie detector.

Women who thought their answers might be seen reported an average of 2.6 sexual partners (all the subjects were college students younger than twenty-five). Those who thought their answers were anonymous reported 3.4 partners, while those who thought their lies would be detected reported an average of 4.4 partners. So, while women admitted to 70 percent more sexual partners when they thought they couldn't fib, the men's answers showed almost no variation. Sex researchers, physicians, and psychologists (and parents) need to remember that women's answers to such questions may depend on when, where, and how the question is asked, as well as who's asking.

If it's true that women's sexuality is much more contextual than most men's, we might need to reconsider a lot of what we think we know about female sexuality. In addition to the distortions created by the age bias we discussed earlier (are twenty-year-olds representative?), how useful are the responses of women answering questions in a cold classroom or laboratory setting? How would our understanding of female sexuality be different if George Clooney distributed the questionnaires by candlelight and collected them after a glass of wine in the Jacuzzi?

Sexologist Lisa Diamond spent over a decade studying the ebb and flow of female desire. In her book *Sexual Fluidity*, she reports that many women see themselves as attracted to specific *people*, rather than to their gender. Women, in Diamond's view, respond so strongly to emotional intimacy that their innate gender orientation can easily be overwhelmed. Chivers agrees: "Women physically don't seem to differentiate between genders in their sex responses, at least heterosexual women don't."

Apparently, many women see the *Mona Lisa* looking back at them from the mirror.

What are the practical effects of this crucial difference in erotic plasticity? To start with, we'd expect to find far more transitory, situational bisexual behavior among women than among men. Various studies of heterosexual couples engaging in group sex or "swinging" agree that it is common for women to have sex with other women in these situations but that men almost never engage with men. Additionally, while we'd be the last to suggest popular culture is a reliable indicator of innate human sexuality, it's probably significant that women kissing women has quickly become accepted as mainstream behavior while depictions of men kissing each other on television or films remains unusual and controversial. Most women presumably wake up the morning after their first same-sex erotic experience more interested in finding some coffee than in conducting a panicked reassessment of their sexual identity. The essence of sexuality for most women seems to include the freedom to change as life changes around them.

There is, after all, a liberating simplicity in Mona Lisa's complexity, which Freud seems to have missed. The answer to his question couldn't be simpler, yet it contains multitudes. What does woman want? It depends.

CHAPTER TWENTY-ONE

The Pervert's Lament

> *Paraphilias are not universally present in human societies, their incidence could be greatly reduced if tolerance and education in sexual matters were more widespread. This is a most important but socially sensitive area of sex research.*
>
> ALAN DIXSON[1]

While many women are freed by their erotic flexibility, men can find themselves trapped by the rigidity of their sexual response, like the male sheep and goats mentioned earlier. Once determined, male eroticism tends to retain its contours throughout life, like concrete that has set. Consequently, the theory of erotic plasticity predicts that paraphilias (abnormal sexual desires and behaviors) should be far more prevalent in men than in women who would presumably be more responsive to social pressures and find it easier to abandon previous turn-ons or ignore unseemly urges. Nearly every source of evidence supports this prediction. Most researchers and therapists agree that these unusual sexual hungers are almost exclusively seen in males, appear to be related to early imprinting, and are difficult, if not impossible, to alter once boyhood impressions have hardened into adult yearnings.

Purely psychological treatment for paraphilias and pedophilia have shown little success. The most effective treatments for the latter tend to be based upon biological approaches (hormone therapy, chemical castration). Once beyond the age of malleability, males seem to be stuck with whatever imprint they've received, latex or leather, S or M, goat or lamb.* If the influences during this "developmental window" are distorting and destructive, a boy may grow into a man with an unalterable, nearly irresistible desire to reenact the same patterns with others. The ritualized, widespread pedophilia in the Catholic Church appears to be a prime example of this process (as does the Church's centuries-long attempt to cover up the issue). Recall Schopenhauer's famous quote, "*Mensch kann tun was er will; er kann aber nicht wollen was er will.*" (One can choose what to do, but not what to want.) Desire, particularly male desire, is notoriously unresponsive to religious dictate, legal retribution, family pressure, self-preservation, or common sense. It does respond to one thing, however: testosterone.

A man who suffered a hormonal disorder that left him with almost no testosterone for four months discussed his experience (anonymously) in a radio interview. Without testosterone, he said, "Everything I identified as being *me* [was lost]. My ambition, my interest in things, my sense of humor, the inflection of my voice. . . . The introduction of testosterone returned everything." Asked whether there had been an upside to being testosterone-free, he said, "There were things that I find offensive about my own personality that were disconnected then. And it was nice to be without them. . . . I approached people with a humility that I had never displayed before." But overall, he was glad to have it back, because, "When you have no testosterone, you have no desire."

Griffin Hansbury, who was born female but underwent a sex change after graduating from college, has another well-informed view of the powers of testosterone. "The world just changes," he said. "The most overwhelming feeling was the incredible increase in libido and change in the way I perceived women." Before the hormone treatments,

* This is not to say that fetishes are paraphilias, merely that they may spring from similar experiences.

Hansbury said, an attractive woman in the street would provoke an internal narrative: "She's attractive. I'd like to meet her." But after the injections, no more narrative. Any attractive quality in a woman, "nice ankles or something," was enough to "flood my mind with aggressive pornographic images, just one after another. . . . Everything I looked at, everything I touched turned to sex." He concluded, "I felt like a monster a lot of the time. It made me understand men. It made me understand adolescent boys a lot."[2]

It doesn't take a sex-change operation to understand that many adolescent boys have a frenzied focus on sex. If you've ever tried to teach a class containing more than a few, or tried to raise one, or recall the turbulent desires yourself, you know the phrase *testosterone poisoning* is not always used ironically. For most adolescent boys, life often seems (and is) violent, hectic, and wild.

Countless studies confirm that testosterone and associated male sex hormones are at high tide from puberty to a man's mid-twenties. Here we have another massive conflict between what society dictates and biology demands. With every voice in a young man's body screaming for SEX NOW, many societies insist he ignore these incessant urgings and channel that energy into other pursuits, ranging from sports to homework to military adventure.[3]

As with other attempts to block the unignorable importations of biology, this one has been a centuries-long disaster. Testosterone levels correlate to the likelihood of a young man (or woman) getting into trouble.[4] In the United States, adolescent males are five times more likely to kill themselves than females are. Among Americans between fifteen and twenty-five, suicide is the third leading cause of death, and teenage boys kill themselves at a rate *double* that of any other demographic group. Adding to the sense that sexual repression underlies this widespread despair, a government study found that homosexual youth are two to three times more likely to attempt suicide than their heterosexual peers.[5]

Well-intentioned websites and presentations rarely, if ever, mention gut-wrenching, identity-clouding sexual frustration as a possible cause of some of this destructive adolescent behavior. Despite the omnipresence of billboards and bus stops featuring semi-naked barely pubescent fashion models, significant parts of American society remain adamantly opposed to any suggestion that sexual activity may commence before the law allows.[6]

In 2003, seventeen-year-old honor student and homecoming king Genarlow Wilson was caught having consensual oral sex with his girlfriend, who had not yet turned sixteen. He was convicted of aggravated child molestation, sentenced to a minimum of ten years in a Georgia prison, and forced to register as a sex offender for life. If Wilson and his girlfriend had just enjoyed good old-fashioned intercourse, as opposed to oral sex, their "crime" would have been a misdemeanor, punishable by a maximum of a year in prison and no sex-offender status.[7]

The previous year, Todd Senters videotaped consensual sex with his girlfriend, who was over the age of consent. No problem, right? Wrong. According to Nebraska state law, although the sex itself was perfectly legal, taping it constituted "manufacturing child pornography." The seventeen-year-old was legally permitted to *have sex*, but images of her doing so are illegal. Go figure.

Adolescents all over the country are getting into serious trouble for *sexting* one another: snapping a risqué photo of themselves with their cell phone and sending it to a friend. Turns out, in many states, these kids can be sent to prison (where sexual abuse is rampant) for photographing their own bodies (manufacturing child pornography) and sharing the photos (distributing child pornography). They're being forced to register as sex offenders despite the fact that they themselves are the "victims" of their own "crimes."[8]

Just Say What?

A 2005 survey of 12,000 adolescents found that those who had pledged to remain abstinent until marriage were *more* likely to have oral and anal sex than other teens, *less* likely to use condoms, and just as likely to contract sexually transmitted diseases as their unapologetically non-abstinent peers. The study's authors found that 88 percent of those who pledged abstinence admitted to failing to keep their pledge.[9]

If our distorted relationship with human sexuality is the source of much of this frustration, confusion, and ignorance, societies with less conflicted views should confirm the causal connection. Developmental neuropsychologist James Prescott found that bodily pleasure and violence seem to have an either/or relationship—the presence of one inhibits development of the other. In 1975, Prescott published a paper in which he argued that "certain sensory experiences during the formative periods of development will create a neuropsychological predisposition for either violence-seeking or pleasure-seeking behaviors later in life." On the level of individual development, this finding seems obvious: adults who abuse children were almost always victims of childhood abuse themselves, and every junkyard owner knows that if you want a mean dog, beat the puppy.

Prescott applied this logic on a cross-cultural level. He performed a meta-analysis of previously gathered data on the amount of physical affection shown to infants (years of breastfeeding, percentage of time in direct physical contact with mother, being fondled and played with by other adults) and overall tolerance for adolescent sexual behavior. After comparing these data with levels of violence within and between societies, Prescott concluded that in all but one of the cultures for which these data were available (forty-eight of forty-nine), "deprivation of body pleasure throughout life—but particularly during the formative periods of infancy, childhood, and adolescence—is very closely related to the amount of warfare and interpersonal violence." Cultures that don't interfere in the physical bonding between mother and child or prohibit the expression of adolescent sexuality show far lower levels of violence—both between individuals and between societies.[10]

While American society twists itself into positions no yoga master could hold (Britney Spears was a vocal virgin while pole-dancing in bikinis on TV?), other societies ritualize and seek to structure adolescent sexuality in positive ways. Mangaian youth are encouraged to have sex with one another, with particular emphasis on the young men learning to control themselves and take pride in the pleasure they can provide a woman. The Muria of central India set up adolescent dormitories (called *ghotuls*), where adolescents are free to sleep together, away from concerned parents. In the ghotul, the young people are encouraged to experiment with different partners, as it's considered unwise to become too attached to a single partner at this phase of life.[11]

If we accept that our species is and always has been optimized for a highly sexual life and that adolescent boys are especially primed for action, why should we be surprised by the explosions of destructive frustration that result from the thwarting of this primal drive?

Kellogg's Guide to Child Abuse

In 1879, Mark Twain gave a speech in which he observed, "Of all the forms of intercourse, [masturbation] has the least to recommend it. As an amusement," he said, "it is too fleeting; as an occupation, it is too wearing; as a public exhibition, there's no money in it."[12] Funny guy, Mark Twain. But there was a seriousness in his humor, as well as courage. As Twain spoke, much of Western culture was waging a bizarre, centuries-long war against any hint of childhood sexuality, including masturbation.

The merciless campaign against masturbation was just one aspect of the West's long struggle against the "sinful" yearnings within human sexuality. We've discussed the so-called witches burned alive for daring to assert or even suggest their eroticism, and doctors like Isaac Baker Brown, who justified barbaric, dangerous surgery as a cure for nascent nymphomania. These were not exceptional cases, as Twain knew. Following the advice of such prominent "experts" as John Harvey Kellogg, many parents of Twain's day subjected their children to brutal physical and mental abuse to stamp out any sign of sexuality. Otherwise reasonable, if confused, people ardently believed that masturbation truly was "the destroying element of civilized society," in the words of the *New Orleans Medical & Surgical Journal.*

Though widely considered to be one of the leading sex educators of his day, Kellogg proudly claimed never to have had intercourse with his wife in over four decades of marriage. But he did require a handsome male orderly to give him an enema every morning—an indulgence his famously high-fiber breakfasts should have made unnecessary. As John Money explains in his study of pseudoscientific anti-sex crusaders, *The Destroying Angel*, Kellogg would probably be diagnosed as a klismaphile today. Klismaphilia is "an anomaly of sexual and erotic functioning traceable to childhood, in which an enema substitutes for regular sexual intercourse. For the klismaphile," writes Money, "putting the penis in the vagina is experienced as hard work, dangerous, and possibly as repulsive."

As a medical doctor, Kellogg claimed the moral authority to instruct parents on the proper sexual education of their children. If you're unfamiliar with the writings of Kellogg and others like him, their gloating disdain for basic human eroticism is chilling and unmistakable. In his best-selling *Plain Facts for Old and Young* (written on his sexless honeymoon in 1888), Kellogg offered parents guidance for dealing with their sons' natural erotic self-exploration in a section entitled "Treatment for Self-Abuse and its Effects." "A remedy which is almost always successful in small boys," he wrote, "is circumcision." He stipulated that, "The operation should be performed by a surgeon *without* administering an anaesthetic, as the brief pain attending the operation will have a salutary effect upon the mind, especially if it be connected with the idea of punishment. . . . [emphasis added]"

If circumcising a struggling, terrified boy without anesthesia wasn't quite what a parent had in mind, Kellogg recommended "the application of one or more silver sutures in such a way as to prevent erection. The prepuce, or foreskin, is drawn forward over the glans, and the needle to which the wire is attached is passed through from one side to the other. After drawing the wire through, the ends are twisted together and cut off close. It is now impossible for an erection to occur. . . ." Parents were assured that sewing their son's penis into its foreskin "acts as a most powerful means of overcoming the disposition to resort to the practice [of masturbation]."[13]

Circumcision remains prevalent in the United States, though varying greatly by region, ranging from about 40 percent of newborns circumcised in western states to about twice that in the Northeast.[14] This widespread procedure, rarely a medical necessity, has its roots in the anti-masturbation campaigns of Kellogg and his like-minded contemporaries. As Money explains, "Neonatal circumcision crept into American delivery rooms in the 1870s and 1880s, not for religious reasons and not for reasons of health or hygiene, as is commonly supposed, but because of the claim that, later in life, it would prevent irritation that would cause the boy to become a masturbator."[15]

Lest you think Kellogg was interested only in the sadistic torture of boys, in the same book he soberly advises the application of carbolic acid to the clitorises of little girls to teach them not to touch themselves. Kellogg and his like-minded contemporaries demonstrate that sexual repression is a "malady that considers itself the remedy," to paraphrase Karl Kraus's dismissal of psychoanalysis.

His smug satisfaction in tormenting children is striking and disturbing, but Kellogg's "no child left alone" policy is anything but unusual or limited to ancient history. The anti-masturbation measures quoted above were published in 1888, but more than eighty years were to pass before the American Medical Association declared, in 1972, "Masturbation is a normal part of adolescent sexual development and requires no medical management." But still, the war continues. As recently as 1994, pediatrician Joycelyn Elders was forced from her post as Surgeon General of the United States for simply asserting that masturbation "is part of human sexuality." The suffering caused by centuries of war on masturbation is beyond calculation. But this we know: all the suffering, every bit of it, was for nothing. *Absolutely nothing.*

John Harvey Kellogg, Anthony Comstock, and Sylvester Graham (inventor of Graham crackers—like corn flakes, a food specifically designed to discourage masturbation) were extreme in their grim campaigns against eroticism, but they weren't considered particularly eccentric at the time.[16] Recall that Darwin probably had had little or no personal sexual experience when he married his first cousin a month before his thirtieth birthday, and that Sigmund Freud—the other towering giant of nineteenth-century sexual theory—was a self-proclaimed thirty-year-old virgin when he married in 1886. No wonder Freud was hesitant sexually. According to biographer Ernest Jones, Freud's father had threatened to cut off young Sigmund's penis if he didn't stop his obsessive masturbating.[17]

The Curse of Calvin Coolidge

> *Last time I tried to make love to my wife nothing was happening. So I said to her, "What's the matter, you can't think of anybody either?"*
>
> RODNEY DANGERFIELD

> *Men don't care what's on TV. They only care what* else *is on TV.*
>
> JERRY SEINFELD

There's a story about President Calvin Coolidge and a chicken farm every evolutionary psychologist knows by heart. It goes like this: The president and his wife were visiting a commercial chicken farm in the 1920s. During the tour, the first lady asked the farmer how he managed to produce so many fertile eggs with only a few roosters. The farmer proudly explained that his roosters happily performed their duty dozens of times each day. "Perhaps you could mention that to the president," replied the first lady. Overhearing the remark, President Coolidge asked the farmer, "Does each cock service the same hen each time?" "Oh no," replied the farmer, "he always changes from one hen to another." "I see," replied the president. "Perhaps you could point that out to Mrs. Coolidge."

Whether the story is historically factual or not, the invigorating effect of a variety of sexual partners has become known as "the Coolidge effect." While there's little doubt that the females of some primate species (including our own) are also intrigued by sexual novelty, the underlying mechanism appears to be different for them. Thus the Coolidge effect generally refers to male mammals, where it's been documented in many species.[18]

But that doesn't mean women's only motivation for sex is relational, as is often argued. Psychologists Joey Sprague and David Quadagno surveyed women from twenty-two to fifty-seven years of age and found that among those under thirty-five, 61 percent of the women said their primary motivation for sex was emotional, rather than physical. But among those *over* thirty-five, only 38 percent claimed their emotional motivations were stronger than the physical hunger for contact.[19] At face value, such results suggest women's motivations change with age. Or one could also argue that this effect could simply reflect women becoming less apologetic as they mature.

First-time travelers to Istanbul, Bali, Gambia, Thailand, or Jamaica may be surprised to see thousands of middle-aged women from Europe and the United States who flock to these places in search of no-strings sexual attention. An estimated eighty thousand women fly to Jamaica looking to "Rent a Rasta" every year.[20] The number of female Japanese visitors to the Thai island resort Phuket jumped from fewer than four thousand in 1990 to ten times that just four years later, outnumbering male Japanese tourists significantly. Chartered jets carrying nothing but Japanese women land in Bangkok every week, if not daily.

In her book *Romance on the Road*, Jeannette Belliveau catalogs dozens of destinations frequented by such women. That this sort of behavior would seem unbelievable and embarrassing to most of the young American women filling out questionnaires for their psychology professors is both result and cause of a more general scientific and cultural blindness to the true contours of female sexuality.

Of course, there are plenty of men looking for sexual variety on the beaches of Thailand as well, but since that just supports the standard narrative, it seems unimportant. Until it becomes very important.

That tiger ain't go crazy; that tiger went tiger! You know when he was really crazy? When he was riding around on a unicycle with a Hitler helmet on!

CHRIS ROCK, *talking about a circus tiger that attacked a trainer*

By temperament, which is the real law of God, many men are goats and can't help committing adultery when they get a chance; whereas there are numbers of men who, by temperament, can keep their purity and let an opportunity go by if the woman lacks in attractiveness.

MARK TWAIN, *Letters from the Earth*

A man we know—we'll call him Phil—could be considered a living icon of male achievement.* In his early forties, handsome, he's been married to Helen, a gorgeous, accomplished physician, for almost twenty years. They have three brilliant, beautiful daughters. Phil and a friend started a small software business in their late twenties and now, fifteen years later, they've both got more money than they'll ever be able to spend. Until recently, Phil lived in a big, beautiful house on a hill overlooking a wooded valley. But Phil's life was, as he puts it, "a disaster waiting to happen."

Disaster struck when Helen discovered the affair he'd been having with a work colleague. Unsurprisingly, she felt deeply betrayed and expressed her outrage by locking him out of the house, refusing even to let him see their children until the lawyers had finished their dismal task. Phil's seemingly perfect life came crashing down around him.

* All names and identifying details have been changed. Please see Author's Note on page 313.

Comedian Chris Rock said, "A man is basically as faithful as his options." Phil's professional success, good looks, and charming personality generated a constant stream of sexual opportunity. Many male readers are probably thinking, "*Of course* he was sleeping with another woman—or two! Come on!" But if you're a woman, you may be thinking, "*Of course* his wife and daughters locked the pig out!"

Is there any way to reconcile these two opposed perspectives on this all-too-common situation? What could possibly motivate so many men who are otherwise demonstrably intelligent, loving, and cautious to risk so much for so little? Everything from the respect of their friends to the love of their children can be lost in the quest for something as transitory and ultimately meaningless as a casual sexual encounter. What are they thinking? We asked Phil.

"At first," he said, "the sex was fantastic. I hadn't felt so alive in years. I thought I was in love with Monica [the other woman]. When I was with her, it was like everything was *stronger*, you know? Food tasted better, colors were richer, I had *so* much more energy. I felt high all the time."

When we asked if the sex he had with Monica was better than it had been with Helen, Phil paused for a long moment. "Actually," he admitted, "now that I think about it, sex with Helen was much better—the best I've ever had, really—at the beginning, you know, those first few years. I mean, with Helen it was never *just* sex. We both knew we wanted to spend our lives together, so there was a depth and, and, well, a *love* and spiritual connection I've never had with anyone else. . . . Even though Helen says she hates me now, I honestly believe we'll always have that connection—even if she won't admit it."

So what happened? "Over the years . . . you know how it is . . . the passion faded and our relationship changed. We became *friends* . . . best friends, but still . . . siblings, almost. It's not her fault. I know this is all my fault, but what can I do?" His eyes tearing up, he said, "It felt like a life-or-death situation. I wanted to feel alive again. I know how ridiculous that sounds, but that's how it felt."

Phil is at a prime age for the so-called *midlife crisis* that seems to hit many men around this stage of their lives. Explanations are easy to come by, ranging from economics (he finally has enough money and status to be attractive to the sort of sexy young women who had ignored him previously) to existential dread (he's coming to terms with his own mortality by lashing out symbolically against his own impending aging and death) to the wife's life cycle (she's nearing menopause, so he's biologically driven toward the fertility of younger women). Each of these may have some measure of truth, but none answers the most pressing question: Why do men have such overwhelming hunger for variety in their sexual partners—not just at midlife, but always?

If the ghost of Calvin Coolidge weren't haunting him, a man would simply buy a DVD or two of his favorite porn actress and watch it over and over the rest of his life. Knowing how the movie ends is hardly going to ruin the experience for him. No, what makes heterosexual men seek a constant stream of *different* women doing the *same* old things is the Coolidge effect. If you've never been to a porn gateway website, you'll be astounded by the variety and specificity of the offerings there: everything from "unshaved Japanese lesbians" to "tattooed redheads" to "overweight older gals." It's a simple, unavoidable truth almost everyone knows to be true but few dare to discuss: variety and change are the necessary spice of the sex life of the human male.

But an intellectual understanding of this aspect of most men's inner reality doesn't make accepting it any easier for many women. Writer and film director Nora Ephron has explored these issues in many of her films, including *Heartburn*, which was based on her own failed marriage. In a 2009 interview she explained how raising two sons had informed her view of men: "Boys are so sweet," she said. "But the problem with men is not whether they're nice or not. It's that it's hard for them at a certain point in their lives to stay true. It just is. It's almost not their fault." But then she added, "It *feels* like it's their fault if you are involved with any of them."[21]

The Perils of Monotomy (Monogamy + Monotony)

The prerequisite for a good marriage, it seems to me, is the license to be unfaithful.

CARL JUNG, *in a letter to Freud dated January 30, 1910*

Remember what Phil said about how sex with his wife had grown overly familiar, how he'd come to feel he and Helen were "siblings, almost"? Interesting word choice. The strongest explanation for the prevalence and intensity of the Coolidge effect among social mammals is that the male drive for sexual variety is evolution's way of avoiding incest. Our species evolved on a sparsely populated planet—never more than a few million and probably fewer than 100,000 of us on Earth for most of our evolutionary past. To avoid the genetic stagnation that would have dragged our ancestors into extinction long ago, males evolved a strong appetite for sexual novelty and a robust aversion to the overly familiar. While this carrot-and-stick mechanism worked well to promote genetic diversity in the prehistoric environment, it's causing lots of problems now. When a couple have been living together for years, when they've become *family*, this ancient anti-incest mechanism can effectively block eroticism for many men, leading to confusion and hurt feelings all around.[22]

Earlier, we discussed how men's testosterone levels recede over the years, but it's not just the passing of time that brings these levels down: monogamy itself seems to drain away a man's testosterone. Married men consistently show lower levels of the hormone than single men of the same age; fathers of young children, even less. Men who are particularly responsive to infants show declines of 30 percent or more right after their child is born. Married men having affairs, however, were found to have higher testosterone levels than those who weren't.[23] Additionally, most of the men having affairs have told researchers they were actually quite happy in their marriages, while only one-third of women having affairs felt that way.[24]

Of course, sharp-thinking readers will point out that these correlations don't imply causation: maybe men with higher levels of testosterone simply seek more affairs. Probably so, but there is good reason to believe that even casual contact with novel, attractive women can have a tonic effect on men's hormonal health. In fact, researcher James Roney and his colleagues found that even a brief chat with an attractive woman raised men's testosterone levels by an average of 14 percent. When these same men spent a few minutes talking with other men, their testosterone level fell by 2 percent.[25]

In the 1960s, anthropologist William Davenport lived among a group of Melanesian islanders who regarded sex as natural and uncomplicated. All the women claimed to be highly orgasmic, with most reporting several orgasms to each of her partner's one. Nevertheless, reported Davenport, "it is assumed that after a few years of marriage, the husband's interest in his wife will begin to pale." Until the recent imposition of colonial laws stopped the practice, these Melanesians avoided monotomy by allowing married men to have young lovers. Rather than being jealous of these concubines, wives regarded them as status symbols, and Davenport claimed that both men and women regarded the loss of this practice the worst result of contact with European culture. "Older men often comment today that without young women to excite them and without the variety once provided by changing concubines, they have become sexually inactive long before their time."[26]

Closer to home, William Masters and Virginia Johnson reported that "loss of coital interest engendered by monotony in a sexual relationship is probably the most constant factor in the loss of an aging male's interest in sexual performance with his partner." They note that this loss of interest can frequently be reversed if the man has a younger lover—even if the lover is not as attractive or sexually skilled as the man's wife. Kinsey concurred, writing, "There seems to be no question but that the human male would be promiscuous in his choice of sexual partners throughout the whole of his life if there were no social restrictions."[27]

For most men and many women, sexual monogamy leads inexorably to monotomy. It's important to understand this process has nothing to do with the attractiveness of the long-term partner or the depth and sincerity of the love felt for him or her. Indeed, quoting Symons, "A man's sexual desire for a woman to whom he is not married is largely the result of her not being his wife."[28] *Novelty itself is the attraction.* Though they're unlikely to admit it, the long-term partners of the sexiest Hollywood starlets are subject to the same psychosexual process. Frustrating? Unfair? Infuriating? Humiliating on both sides? Yes, yes, yes, and yes. But still, true.

What to do about it? Most modern couples aren't as flexible about tolerating a variety of sexual partners as the Melanesians and many of the societies we surveyed in earlier chapters. After reviewing the broad literature on Western marriage, sociologist Jessie Bernard argued in the early 1970s that increasing men's opportunities for sexually novel partners was one of the most important social changes required in Western societies to promote marital happiness.[29] But this hasn't happened yet and seems even less likely now, almost four decades later. Maybe this is why some twenty million American marriages can be categorized as no-sex or low-sex due to the *man's* loss of sexual interest. According to the authors of *He's Just Not Up for It Anymore*, 15 to 20 percent of American couples have sex fewer than ten times per year. They note that the absence of sexual desire is the most common sexual problem in the country.[30] Combine these dismal numbers with the 50 percent of all marriages that end in divorce, and it's clear that modern marriage is suffering a soft-core meltdown.

In *The Evolution of Human Sexuality*, the ever-quotable Donald Symons pointed out that Western societies have tried every trick in the book to change this aspect of male sexuality, but all have failed miserably: "Human males seem to be so constituted that they resist learning not to desire variety," he wrote, "despite impediments such as Christianity and the doctrine of sin; Judaism and the doctrine of mensch; social science and the doctrines of repressed homosexuality and psychosexual immaturity; evolutionary theories of monogamous pair-bonding; cultural and legal traditions that support and glorify monogamy."[31] Need we supplement Symons's thoughts with a list of specific examples of men (presidents, governors, senators, athletes, musicians) who have squandered family and fortune, power and prestige—all for an encounter with a woman whose principal attraction was her novelty? Need we remind female readers of the men in their past who seemed so smitten at first, but mysteriously stopped calling once the thrill of novelty had faded?

A Few More Reasons I Need Somebody New (Just Like You)

> *Making love with a woman and sleeping with a woman are two separate passions, not merely different but opposite. Love does not make itself felt in the desire for copulation (a desire that extends to an infinite number of women) but in the desire for shared sleep (a desire limited to one woman).*
>
> MILAN KUNDERA, *The Unbearable Lightness of Being*

Remember how Phil said he felt "high" when he was with his new lover? "Colors were richer, food tasted better." There's a reason for this intensification of sensation, but it's not love. As their testosterone levels decline with age, many men experience a diminishment of energy and libido, an intangible distance from the basic pleasures of life. Most attribute this blurred distance to stress, lack of sleep, or too much responsibility, or they just chalk it up to the passage of time. True enough, but some of this numbing could be due to ebbing testosterone levels. Recall the man who had no testosterone for a while. He felt he'd lost "everything I identify as being *me*." His ambition, passion for life, sense of humor . . . all gone. Until testosterone brought it all back. Without the testosterone, he said, "you have no desire."

Phil thought he was in love. *Of course he did.* As suggested above, one of the few things that reliably revives a male's sagging testosterone levels is a novel lover.[32] So he felt all the things we associate with love: renewed vitality, a new depth and intensification—a giddy thrill at being alive. How easily we mistake this potent mix of feelings for "love." But a hormonal response to novelty isn't love.

How many men have mistaken this hormone high for a life-altering spiritual union? How many women have been blindsided by a good man's seemingly inexplicable betrayal? How many families have been ripped apart because middle-aged men misinterpreted the surge of vitality and energy resulting from a novel sexual partner as love for a soul mate—or convinced themselves they were in love to justify what felt like a life-affirming necessity? And how many of these men then found themselves isolated, shamed, and devastated when the curse of Coolidge returned after a few months or years to reveal that the now-familiar partner was not, in fact, the true source of those feelings after all? No one knows the number, but it's a big one.

This common situation is heavy with tragedy, but one of the most painful aspects may be that many of these men will realize that the woman they left behind was a far better match than the one they left her for. Once the transitory thrill passes, these men are left once again with the realities of what makes a relationship work over the long run: respect, admiration, convergent interests, good conversation, sense of humor, and so on. A marriage built upon sexual passion alone has as much chance of enduring as a house built on winter ice. Only by arriving at a more nuanced understanding of the nature of human sexuality will we learn to make smarter decisions about our long-term commitments. But this understanding requires us to face some uncomfortable facts.

Like many men in the same situation, Phil said he felt as if he faced a "life or death" decision. Maybe he did. Researchers have found that men with lower levels of testosterone are more than four times as likely to suffer from clinical depression, fatal heart attacks, and cancer when compared to other men their age with higher testosterone levels. They are also more likely to develop Alzheimer's disease and other forms of dementia, and have a far greater risk of dying prematurely from any cause (ranging from 88 to 250 percent higher, depending on the study).[33]

If it's true that most men are constituted, by millions of years of evolution, to need occasional novel partners to maintain an active and vital sexuality throughout their lives, then what are we saying to men when we demand lifetime sexual monogamy? Must they choose between familial love and long-term sexual fulfillment? Most men don't fully appreciate the conflict between the demands of society and those of their own biology until they've been married for years—plenty of time for life to have grown very complicated, with children, joint property, mutual friends, and the sort of love and friendship only shared history can bring. When they arrive at the crisis point, where domesticity and declining testosterone levels have drained the color from life, what to do?

The options most men see before them seem to be:

1. Lie and try not to get caught. While this option may be the most commonly chosen, it may also be the worst. How many men think they have an "unspoken agreement" with their wife that, as long as she doesn't find out about it, it's okay for him to have a casual relationship on the side? This is like saying you have an unspoken agreement with the police that it's okay to drive drunk—as long as they don't catch you. Even if there *is* some understanding along these lines, any lawyer will tell you that unspoken agreements are the worst possible foundation for any long-term partnership.

   A. Gentlemen, you are going to get caught sooner or later (probably sooner). You have as much chance of getting away with this as a dog has of following a cat up a tree. Ain't gonna happen. One reason: most women's sense of smell is significantly better than most men's, so there's probably going to be evidence you can't even sense, but that she'll pick up on. Need we even mention the much-vaunted powers of female intuition?

B. This requires you to lie to your partner in life. To deceive the mother of your children, the person you were hoping to grow old with. Is this really who you are? Is this the man she chose to share her life with?

2. Give up on having sex with anyone other than your wife for the rest of your life. Maybe resort to porn and Prozac.

A. Antidepressants are the most prescribed drug in the United States, with 118 million prescriptions written in 2005 alone. One of the most prominent side effects of these drugs is the dampening of libido, so maybe the whole issue will just fade away—chemical castration. If not, there's always Viagra, with well over a *billion* tablets doled out in the decade since it was introduced in 1998. But Viagra creates *blood flow*, not *desire*. Now men can fake sexual interest too. Progress?

B. It's not the same, is it? And isn't there something humiliating (not to say emasculating) about sneaking off at night to look at porn on your computer? This course often leads to serious anger and resentment that can destroy a relationship.

3. Serial monogamy: divorce and start over. This option seems to be the "honest" approach recommended by most experts—including many relationship counselors.

A. Serial monogamy is a symptomatic response to the issues posed by the conflict between what society dictates and what biology demands. It solves nothing in terms of snowballing male (and thus, female) sexual frustration in long-term sexually monogamous relationships.

B. Though often presented as the *honorable* response to the conundrum, the serial monogamy cop-out has led directly to the current epidemic of broken homes and single-parent families. How is it "adult" to inflict emotional trauma on

> our children because we're unable to face the truth about sex? Susan Squire, author of *I Don't: A Contrarian History of Marriage*, asks: "Why does society consider it more moral for you to break up a marriage, go through a divorce, disrupt your children's lives maybe forever, just to be able to fuck someone with whom the fucking is going to get just as boring as it was with the first person before long?"[34] A man who pursues long-term happiness by leaving behind a string of hurt, embittered women and emotionally wounded children is little more than a dog chasing tail—his own.

And if you're a woman whose husband is "cheating," your options are no better: pretend you don't notice what's going on, go out and have your own revenge affair (even if you don't feel like it), or destroy your own family and marriage by calling in the lawyers. These are all losing scenarios.

Even the term we use to describe this betrayal of self and family, "cheating," echoes the standard narrative of human sexuality in its implication that marriage is a game that one player can win at the expense of the other. The woman who "tricks" a man into supporting children he *thinks* are his has, according to this model, cheated—and won. Another big winner, according to the standard narrative, is the "baby-daddy" who manages to impregnate a string of women who then raise his children while he's already on to his next conquest. But in any true *partnership*—married or not—cheating cannot lead to any sort of victory. It's win-win or everybody loses.

CHAPTER TWENTY-TWO

Confronting the Sky Together

Love is not breathlessness, it is not excitement, it is not the promulgation of promises of eternal passion. That is just being "in love" which any of us can convince ourselves we are. Love itself is what is left over when being in love has burned away...

—Louis de Bernières, *Correlli's Mandolin*

There is a cost . . . for the society that insists on conformity to a particular range of heterosexual practices. We believe that cultures can be rationally designed. We can teach and reward and coerce. But in so doing we must also consider the price of each culture, measured in the time and energy required for training and enforcement and in the less tangible currency of human happiness that must be spent to circumvent our innate predispositions.

E. O. Wilson[1]

So now what? Having written this whole book about sex, we'd like to confusingly suggest that most of us take sex way too seriously: when it's just sex, that's all it is. In such cases, it's not love. Or sin. Or pathology. Or a good reason to destroy an otherwise happy family.

Like the Victorians, most contemporary Western societies inflate the inherent value of sex by restricting supply ("Good girls don't") and inflating demand (*Girls Gone Wild*). This process leads to a distorted vision of just how important sex actually is. Yes, sex is essential, but it's not something that must *always* be taken so seriously. Think of food, water, oxygen, shelter, and all the other elements of life crucial to survival and happiness but that don't figure in our day-to-day thinking unless they become unavailable. A reasonable relaxation of moralistic social codes making sexual satisfaction more easily available would also make it less problematic.

This appears to be the overall trajectory of history. While many are perplexed and disturbed by the "hooking up" culture, the *sexting* of racy images back and forth, full recognition of all legal rights for gay male and lesbian couples, and so on, there's not much they can do to stop any of it for long. In terms of sexuality, history appears to be flowing back toward a hunter-gatherer casualness. If so, future generations may suffer fewer pathological manifestations of sexual frustration and unnecessarily fractured families. Concerning the Siriono, with whom he lived, Holmberg writes, "The Siriono rarely, if ever, lack for sexual partners. Whenever the sex drive is up, there is almost always an available partner willing to reduce it. . . . Sex anxiety seems to be remarkably low in Siriono society. Such manifestations as excessive indulgence, continence, or sex dreams and fantasies are rarely encountered."[2]

How would it feel to live in such a world? Well, we all know how it feels to live in this one. Apart from death itself, what causes as much human misery as the ongoing demise of marriage? In 2008, almost 40 percent of the mothers who gave birth in the United States were unmarried. This matters. As reported recently in *Time*, "On every single significant outcome related to short-term well-being and long-term success, children from intact, two-parent families outperform those from single-parent households. Longevity, drug abuse, school performance and dropout rates, teen pregnancy, criminal behavior and incarceration . . . in all cases, the kids living with both parents drastically outperform the others."[3]

"Love is an ideal thing, marriage a real thing," observed German philosopher Johann Wolfgang von Goethe. "A confusion of the real with the ideal never goes unpunished." Indeed. By insisting upon an ideal vision of marriage founded upon a lifetime of sexual fidelity to one person—a vision most of us eventually learn is highly unrealistic, we invite punishment upon ourselves, upon each other, and upon our children.

"The French are much more comfortable with the idea that their affair partner is just that—an affair partner," writes Pamela Druckerman in her cross-cultural look at infidelity, *Lust in Translation*. Understanding that love and sex are different things, Druckerman says the French feel less need to "complain about their marriage to legitimize the affair in the first place." But she found that Americans and British couples seemed to be reading from an entirely different script. "An affair, even a one-night stand, means a marriage is over," Druckerman observed. "I spoke to women who, on discovering that their husbands had cheated, immediately packed a bag and left, because 'that's what you do.' Not because that's what they *wanted* to do—they just thought that was the rule. They didn't even seem to realize there were other options. . . . I mean, really, like they're reading from a script!"[4]

Psychologist Julian Jaynes described the commingling of terror and exhilaration one experiences upon realizing that things are not as they'd seemed: "There is an awkward moment at the top of a Ferris wheel when, having come up the inside curvature, where we are facing into a firm structure of confident girders, suddenly that structure disappears, and we are thrust out into the sky for the outward curve down."[5] This is the moment too many couples struggle in vain to avoid or ignore—even to the point of choosing bitter divorce and fractured family over the daunting task of confronting the sky together, with all the "confident girders" behind them in the past.

The false expectations we hold about ourselves, each other, and human sexuality do us serious, lasting harm. As author and sex advice columnist Dan Savage explains, "The expectation of lifelong monogamy places an incredible strain on a marriage. But our concept of love and marriage has as its foundation not only the expectation of monogamy but the idea that where there's love, monogamy should be easy and joyful."[6]

To be sure, toe-curlingly passionate sex can be an important part of marital intimacy, but it is a grave mistake to think it's the essence of long-term intimacy. Like every other kind of hunger, sexual desire tends to be smothered by its satisfaction. Squire says that thinking of marriage as an enduring romance is unrealistic: "It's not like you want to rip your clothes off with somebody that you're sleeping with for the thousandth time. We should know going into it that the nature of love and sex changes from what it begins as, and that a great love affair doesn't necessarily make a great marriage."[7] High-libido sex can just as easily be an expression of the utter absence of intimacy: consider the notorious one-night stand, the prostitute, basic physical release.

Couples might find that the only route to preserving or rediscovering intensity reminiscent of their early days and nights requires confronting the open, uncertain sky together. They may find themselves having their most meaningful, intimate conversations if they dare to talk about the true nature of their feelings. We don't mean to suggest these will be easy conversations. They won't be. There are zones where it's always going to be difficult for men and women to understand one another, and sexual desire is one of them. Many women will find it difficult to accept that men can so easily dissociate sexual pleasure from emotional intimacy, just as many men will struggle to understand why these two obviously separate (to them) issues are often so intertwined for many women.

But with trust, we can strive to accept even what we cannot understand. One of the most important hopes we have for this book is to provoke the sorts of conversations that make it a bit easier for couples to make their way across this difficult emotional terrain together, with a deeper, less judgmental understanding of the ancient roots of these inconvenient feelings and a more informed, mature approach to dealing with them. Other than that, we really have little helpful advice to offer. Every relationship is a constantly changing world that requires specific attention. Other than warning you to be wary of those who offer one-size-fits-all relationship advice, our best counsel echoes that of Polonius to Laertes (in *Hamlet*): "To thine own self be true, and it must follow, as the night the day, thou canst not then be false to any man [or woman]."

Still, it will take more than a deeper understanding of ourselves and each other to fully address the many issues raised by a more relaxed and tolerant approach to fidelity. "The people I feel sorry for are the ones who don't even realize they have any other choices beyond the traditional options society presents," says Scott, who is a member of a long-term triad relationship with Terisa (a woman) who is also involved with Larry (whom Scott introduced to Terisa). While such three- or four-person committed relationships have, by necessity, flown under the radar until recently, so-called polyamorous families are thought to number about half a million in the United States, according to an article in Newsweek.[8] Although Helen Fisher thinks people involved in such configurations are "fighting Mother Nature" by trying to confront their insecurities and jealousy head-on, there is plenty of evidence that, for the right people, such arrangements can work out very well for all concerned—even the kids.

As Sarah Hrdy reminds us, conventional couples struggling to raise a family in isolation might be the ones fighting Mother Nature: "Since Darwin," she writes, "we have assumed that humans evolved in families where a mother relied on one male to help her rear her young in a nuclear family; yet . . . the diversity of human family arrangements . . . is better predicted by assuming that our ancestors evolved as cooperative breeders."[9] From our perspective, people like Scott, Larry, and Terisa appear to be trying to replicate ancient human socio-sexual configurations. As we've seen, from a child's perspective, having more than two stable, loving adults around can be enriching, whether in Africa, the Amazon, China, or suburban Colorado. Laird Harrison recently wrote about his experience growing up in a house his biological parents shared with another couple and their children. He recalled, "The communal household enjoyed a kind of camaraderie I have never felt since. . . . I swapped books with my stepsisters, listened in awe to their stories of crushes, exchanged tips on teachers. Their father imparted his love of great music and their mother her passion for cooking. A sort of bond formed among the 10 of us."[10]

Everybody Out of the Closet

> *An era can be considered over when its basic illusions have been exhausted.*
>
> ARTHUR MILLER

Much of recent history can be seen as waves of tolerance and acceptance breaking against the rocky headlands of rigid social structures. Though it can seem to take almost forever, the waves always win in the end, reducing immobile rock to shifting sand. The twentieth century saw the headlands beginning to crumble under surges of anti-slavery movements, women's rights, racial equality, and, more recently, the steadily growing acceptance of the rights of gay, lesbian, transgender, and bisexual people.

Author Andrew Sullivan described his experience growing up both gay and Catholic as "difficult to the point of agony. I saw in my own life and those of countless others," Sullivan recalled, "that the suppression of these core emotions and the denial of their resolution in love always *always* leads to personal distortion and compulsion and loss of perspective. Forcing . . . people into molds they do not fit helps no one," Sullivan wrote. "It robs them of dignity and self-worth and the capacity for healthy relationships. It wrecks family, twists Christianity, violates humanity. It must end."[11] Sullivan's comments were provoked by the twisted collapse of the publicly homophobic but privately homosexual televangelist Ted Haggard, but he could have been speaking for anyone who doesn't fit the socially sanctioned mold of his or her day.

And who *does* fit this mold? Yes, self-hating gay televangelists and politicians need to come out of the closet, but so does everyone else.

It won't be easy. It's never easy to stand up to shame-fueled anger. Historian Robert S. McElvaine previews some of the shrill denunciation awaiting those who may dare to wander from the monogamous fold, declaring, "Free love is likely to degenerate into 'free hate.' Since loving everybody is a biological impossibility, the attempt to do so [becomes] 'otherization,' and the hatred that goes with it."[12] Like McElvaine, many relationship counselors seem both terrified by and ignorant of nonstandard marital relationships of any kind. Esther Perel, author of *Mating in Captivity*, quotes a family therapist she knows (and respects) stating unequivocally, "Open marriage doesn't work. Thinking you can do it is totally naïve. We tried it in the seventies and it was a disaster."[13]

Maybe, but such therapists might want to delve a little deeper before reflexively dismissing alternatives to conventional marriage. Asked to imagine the first swingers in modern American history, most people probably picture hairy hippies in headbands lolling about on waterbeds in free-love communes under posters of Che Guevara and Jimi Hendrix, Jefferson Airplane on the hi-fi. But be cool, Daddy-O, 'cause the truth is gonna blow your mind.

It seems that the original modern American swingers were crew-cut World War II air force pilots and their wives. Like elite warriors everywhere, these "top guns" often developed strong bonds with one another, perhaps because they suffered the highest casualty rate of any branch of the military. According to journalist Terry Gould, "key parties," like those later dramatized in the 1997 film *The Ice Storm*, originated on these military bases in the 1940s, where elite pilots and their wives intermingled sexually with one another before the men flew off toward Japanese antiaircraft fire.

Gould, author of *The Lifestyle*, a cultural history of the swinging movement in the United States, interviewed two researchers who'd written about this Air Force ritual. Joan and Dwight Dixon explained to Gould that these warriors and their wives "shared each other as a kind of tribal bonding ritual, with a tacit understanding that the two thirds of husbands who survived would look after the widows."* The practice continued after the war ended and by the late 1940s, "military installations from Maine to Texas and California to Washington had thriving swing clubs," writes Gould. By the end of the Korean War, in 1953, the clubs "had spread from the air bases to the surrounding suburbs among straight, white-collar professionals."[14]

Are we to believe that these fighter pilots and their wives were "naïve"?

It's true that many high-profile American forays into alternative sexuality in the 1970s ended in chaos and hurt feelings, but what does that prove? Americans also tried, and failed, to reduce their reliance on foreign oil in the 1970s. By this logic, it would be "naïve" to ever try again. Besides, in intimate matters, discretion and success tend to go together, so no one really has any idea how many couples *succeeded* in finding their own unconventional understandings by experimenting with low-key alternatives to standard, off-the-shelf monogamy.[15]

What isn't debatable is that conventional marriage is a full-blown di-

* Recall Beckerman's description of mate-sharing in the Amazon: "You know that if you die, there's some other man who has a residual obligation to care for at least one of your children. So looking the other way or even giving your blessing when your wife takes a lover is the only insurance you can buy."

saster for millions of men, women, and children right now. Conventional till-death-(or infidelity, or boredom)-do-us-part marriage is a failure. Emotionally, economically, psychologically, and sexually, it just doesn't work over the long term for too many couples. Yet while few mainstream therapists would contemplate trying to convince a gay man or lesbian to "grow up, get real, and just stop being gay," these days, when it comes to unconventional approaches to heterosexual marriage, Perel points out, "Sexual boundaries are one of the few areas where therapists seem to mirror the dominant culture. Monogamy," she writes, "is the norm, and sexual fidelity is considered to be mature, committed, and realistic." Forget about negotiating alternatives: "Nonmonogamy, even consensual nonmonogamy, is suspect." The notion that it might be possible to love one person while being sexual with another "makes us shudder," and conjures "images of chaos: promiscuity, orgies, debauchery."[16]

Couples who turn to a therapist hoping for guidance on ways to loosen—but not break—the bonds of standard monogamy are likely to be offered little but defensive condemnation and stilted bromides like this advice from an evolutionary psychology based self-help book called *Mean Genes*: "The temptations we all face are deeply ingrained in the genes of our hearts and minds . . . [but] as long as we remain interesting dynamos, there will be no conflict between monogamy and our infidelity-promoting mean genes."[17] Interesting dynamos? No conflict? Sure. Tell that to Mrs. Coolidge.

Perel is the rare therapist willing to publicly consider the possibility that heterosexual couples might find alternative arrangements that can work well for them—even if they find themselves outside the bounds of what mainstream society approves. She writes, "It's been my experience that couples who negotiate sexual boundaries . . . are no less committed than those who keep the gates closed. In fact, it is their desire to make the relationship stronger that leads them to explore other models of long-term love."[18]

There are an infinite number of ways to adapt a flexible and loving partnership to our ancient appetites. Despite what most mainstream therapists claim, for example, couples with "open marriages" generally rate their overall satisfaction (with both their relationship and with life in general) significantly higher than those in conventional marriages do.[19] Polyamorists have found ways to incorporate additional relationships into their lives without lying to one another and destroying their primary partnership. Like many gay male couples, these people recognize that additional relationships need not be taken as indictments of anyone. Dossie Easton and Catherine Liszt, authors of *The Ethical Slut*, write, "It is cruel and insensitive to interpret an affair as a symptom of sickness in the relationship, as it leaves the 'cheated-on' partner—who may already be feeling insecure—to wonder what is wrong with him. . . . Many people have sex outside their primary relationships for reasons that have nothing to do with any inadequacy in their partner or in the relationship."[20]

Despite centuries of religious and scientific propaganda, the basic illusions underpinning the supposed "naturalness" of the conventional nuclear family are clearly exhausted. This collapse has left many of us isolated and unfulfilled. Blind insistence and well-intentioned inquisitions have failed to turn the tide, and show no signs of future success. Rather than endless War Between the Sexes, or rigid adherence to a notion of the human family that was never true to begin with, we need to seek peace with the truths of human sexuality. Maybe this means improvising new familial configurations. Perhaps it will require more community assistance for single mothers and their children. Or maybe it just means we must learn to adjust our expectations concerning sexual fidelity. But this we know: vehement denial, inflexible religious or legislative dictate, and medieval stoning rituals in the desert have all proved powerless against our prehistoric predilections.

In 1988, Roy Romer, then governor of Colorado, faced a feeding-frenzy of questions about his long-running extramarital affair that had become publicly known. Romer did what few public figures have dared. In the spirit of the Yucatán, he refused to accept the premise underlying the intrusive questions: that his extramarital relationship was a betrayal of his wife and family. Instead, he called an extraordinary press conference where he pointed out that his wife of forty-five years had known about and accepted the relationship all along. Romer confronted the tittering reporters with "life as it really happens." "What is fidelity?" he asked the suddenly silent gaggle of reporters. "Fidelity is what kind of openness you have. What kind of trust you have, which is based on truth and openness. And so, in my own family, we've discussed that at some length and we've tried to arrive at an understanding of what our feelings are, what our needs are, and work it out with *that* kind of fidelity."[21]

The Marriage of the Sun and the Moon

In a sky swarming with uncountable stars, clouds endlessly flowing, and planets wandering, always and forever there has been just one moon and one sun. To our ancestors, these two mysterious bodies reflected the female and the male essences. From Iceland to Tierra del Fuego, people attributed the Sun's constancy and power to his masculinity; the Moon's changeability, unspeakable beauty, and monthly cycles were signs of her femininity.

To human eyes turned toward the sky 100,000 years ago, they appeared identical in size, as they do to our eyes today. In a total solar eclipse, the disc of the moon fits so precisely over that of the sun that the naked eye can see solar flares leaping into space from behind.

But while they *appear* precisely the same size to terrestrial observers, scientists long ago determined that the true diameter of the sun is about *four hundred times* that of the moon. Yet incredibly, the sun's distance from Earth is roughly four hundred times that of the moon's, thus bringing them into unlikely balance when viewed from the only planet with anyone around to notice.[22]

Some will say, "Interesting coincidence." Others will wonder whether there isn't an extraordinary message contained in this celestial convergence of difference and similarity, intimacy and distance, rhythmic constancy and cyclical change. Like our distant ancestors, we watch the eternal dance of our sun and our moon, looking for clues to the nature of man and woman, masculine and feminine here at home.

Luc Viatour/www.lucnix.be

AUTHORS' NOTE

The material in Chapter 21 featuring "philandering Phil" struck some readers of previous editions of this book as imbalanced and even hypocritical in light of all we've said about the importance of sexual satisfaction for both women and men. "Why only talk about having an affair from a man's perspective," we've been asked, "when the rest of your book is so balanced and supportive of women's sexuality?" That's a fair and direct question for which we can offer only unfair and indirect answers.

First, many men report that they had affairs simply because opportunities arose, while women—for whom opportunities are often more plentiful—tend to report a more complex confluence of motivations. For example, when Shirley Glass and Thomas White anonymously interviewed three hundred men and women about their extramarital affairs, they found that men tended to see their affairs as more sexual while women were motivated more by emotional considerations and reported greater levels of dissatisfaction with their marriages. These findings have been echoed repeatedly in other research.

Second, as we've discussed in previous chapters, women's libidinous motivations tend to be far more fluid, and thus harder to discuss adequately, than men's. Recall that women are more likely to engage in extramarital sex when they're ovulating, for example, and are less likely to use birth control than at other points in their menstrual cycle. A woman in her forties may well approach a "friends with benefits" situation completely differently than she would have two decades earlier, for reasons relating both to hormonal levels and life experience.

In addition to these internal factors, women tend to be more responsive to external conditions. (Are the kids grown and out of the house? Is she financially independent? What would her friends and family say? Does she suspect that he's having an affair?) Men—even highly intelligent, otherwise cautious and calculating men—often blunder into these situations blinded by something that doesn't seem to render women quite so helpless.

Of course, none of this is definitive or universal. Whatever generalizations we make about motivations will be belied by many, many exceptions among both men and women. Every person is a world and every relationship a universe. Nothing we say here is meant to simplify or minimize anyone's experience, male or female.

Our purpose is merely to briefly explore how some of the theories we've discussed play out in many modern lives by looking at the scenario married couples confront most frequently: the middle-aged man who strays. A similar assessment of women's motivations and experiences of extramarital affairs would require far more space than we have. Plus, we actually know "Phil," who was willing to discuss his experience with us. If we know any women who are having affairs as we write this, they've chosen not to share their secret with us, perhaps wisely.

ACKNOWLEDGMENTS

They say publishing a book is like having a baby, but it takes longer and hurts more. Appropriately, this "baby" has many parents. There would be no *Sex at Dawn* without the insight, encouragement, and patience of our families, especially Frank, Julie, and Beth Ryan, Joana and Manel Ruas, Alzira Remane, Celestino Almeida, and Danial Jethá.

Stephan Lang (whose name we inexcusably misspelled in the hardcover version of this book) and Henriette Klauser were incredibly generous in helping us put together a convincing book proposal. Our agent, Melissa Flashman, spent countless hours guiding us through the transition from proposal to manuscript. Unlike most agents, she kept reading and offering wise counsel throughout the entire publishing process, for which we are sincerely grateful. Many thanks to Ben Loehnen, our editor at HarperCollins, who believed in the book from the get-go (even while no doubt discreetly disagreeing with some of its content), and assistant editor Matthew Inman, for his rapid-response professionalism. Lisa Wolff did a first-class copyedit, catching more than a few potentially embarrassing mistakes. Those that snuck through or that we slipped in later are nobody's fault but our own.

Frank Ryan (WBE), Stanton Peele, Stanley Krippner, Julie Holland, Britt Winston, and Steve Mason masochistically read and re-read early drafts of the entire messy manuscript. Their comments were sadistically honest, which is exactly what we needed. In addition to their crucial scholarship, Robert Sapolsky, Todd Shackelford, Helen Fisher, Daniel Moses, and Frans de Waal contributed scarce free time to review parts of the manuscript.

Finally, we thank the following people (in random order) for the many kinds of support and encouragement they've given us: Michael and Mireille Lang, Brian and Crosby O'Hare, Marta Cervera, Alejandra Peña, Dorothianne Henne, Naomi and Don Norwood, Octavi de Daniel, Adam Mendelson, Richard Schweid, David Darnell, Señor Manolo Reyes, Matt Dondet, Mark Plummer, Cybele Tom, Sean Doyle, Santiago Suso, Victoria Ribera, Antonio Berruezo, Eric Patterson, Don Cooper, Martijn van Duivendijk, Peggy and Raul Rossel, Nacho and Leo Valls-Jové, Celine Salvans, Carmen Palomar Lopez, Anamargarita Otero-Robertson, Viram, Voodoo, Maria da Luz Venâncio Guerreiro, João Alves Falcato, Mario Simões, Steve Taylor, Vince and Carrie Stamper, Susie Bright, Jacqui Deegan, and, of course, Dan, Terry, and D.J.

NOTES

Please visit sexatdawn.com for the latest news, further discussion, and updates on the issues raised in this book, or to contact the authors.

Introduction

1. Maybe as recently as 4.5 million years ago. For a recent review of the genetic evidence, see Siepel (2009).
2. de Waal (1998), p. 5.
3. Some of these numbers are reported in McNeil et al. (2006) and Yoder et al. (2005). The hundred billion figure comes from http://www.latimes.com/news/nationworld/nation/la-fg-vienna-porn25-2009mar25,0,7189584.story.
4. See "Yes, dear. Tonight again." Ralph Gardner, Jr. *The New York Times* (June 9, 2008): http://www.nytimes.com/2008/06/09/arts/09iht-08nights.13568273.html?_r=1
5. Full disclosure: Murdoch also owns HarperCollins, the publisher of this book.
6. Diamond (1987).
7. Such relationships would have been among many group-identity-boosting techniques, including participation in group bonding rituals still common to shamanistic religions characteristic of foraging people. Interestingly, such collective-identity-affirming rituals are often accompanied by music (which—like orgasm—releases oxytocin, the hormone most associated with forming emotional bonds). See Levitin (2009) for more on music and social identity.
8. The precise timing of this shift has recently been called into question. See White and Lovejoy (2009).
9. For more on the sharing-based economies of foragers, see Sahlins (1972), Hawkes (1993), Gowdy (1998), Boehm (1999), or Michael Finkel's *National Geographic* article on the Hadza, available here: http://ngm.nationalgeographic.com/2009/12/hadza/finkel-text.
10. Mithen (2007), p. 705.

11. Taylor (1996), pp. 142–143. Taylor's book is an excellent archaeological account of human sexual origins.

Part I: On the Origin of the Specious

Chapter 1: Remember the Yucatán!

1. This account comes from Todorov (1984), but Todorov's version of events is not universally accepted. See http://www.yucatantoday.com/culture/esp-yucatan-name.htm, for example, for a review of other etymologies (in Spanish).
2. From the FDA's Macroanalytical Procedures Manual—Spice Methods. Accessed online at: http://www.fda.gov/Food/ScienceResearch/LaboratoryMethods/MacroanalyticalProceduresManualMPM/ucm084394.htm.

Chapter 2: What Darwin Didn't Know About Sex

1. Originally published in *Daedalus*, Spring 2007. Article can be found here: http://www.redorbit.com/news/science/931165/challenging_darwins_theory_of_sexual_selection/index.html. For more of her uniquely informed view of sexual diversity in nature, see Roughgarden (2004). For her deconstruction of self-interest as the engine of natural and sexual selection, see Roughgarden (2009). For more on homosexuality in the animal world, see Bagemihl (1999).
2. http://www.advicegoddess.com/ag-column-archives/2006/05.
3. Not everyone would agree, of course. When Darwin's brother Erasmus first read the book, he found Charles's reasoning so compelling that he wasn't bothered by the lack of evidence, writing, "If the facts won't fit in, why so much the worse for the facts is my feeling."

   For a thorough (but reader-friendly) look at how Darwin's Victorianism affected his own and subsequent science, see Hrdy (1996).
4. Darwin (1871/2007), p. 362.
5. Pinker (2002), p. 253.
6. Fowles (1969), pp. 211-212.
7. Houghton (1957). Quoted in Wright (1994), p. 224.
8. Quoted in Richards (1979), p. 1244.
9. Writing in *Scientific American Online* (February 2005, p. 30), science historian Londa Schiebinger explains: "Erasmus Darwin . . . did not limit sexual relations to the bonds of holy matrimony. In his *Loves of*

the Plants (1789), Darwin's plants freely expressed every imaginable form of heterosexual union. The fair *Collinsonia*, sighing with sweet concern, satisfied the love of two brothers by turns. The *Meadia*—an ordinary cowslip—bowed with 'wanton air,' rolled her dark eyes and waved her golden hair as she gratified each of her five beaux. . . . Darwin may well have been using the cover of botany to propagandize for the free love he practiced after the death of his first wife."

10. From Hrdy (1999b).
11. Raverat (1991).
12. Desmond and Moore (1994), p. 257. Also, see Wright (1994) for excellent insights into Darwin's thought process and family life.
13. Levine (1996) first used the term *Flintstonization. The Flintstones* occupies a unique place in American cultural history. It was the first prime-time animated series for adults, the first prime-time animated series to last more than two seasons (not matched until *The Simpsons* in 1992), and the first animated program to show a man and woman in bed together.
14. Lovejoy (1981).
15. Fisher (1992), p. 72.
16. Ridley (2006), p. 35.
17. See, for example, Steven Pinker's assertion that human societies have become progressively more peaceful through the generations (discussed in detail in Chapter 13).
18. Wilson (1978), pp. 1–2.
19. A view Steven Pinker resuscitated decades later, long after more nuanced positions had become prevalent.
20. See, for example, Thornhill and Palmer (2000).
21. "A Treatise on the Tyranny of Two," *New York Times Magazine*, October 14, 2001. You can read the essay online at http://www.nytimes.com/2001/10/14/magazine/14AGAINSTLOVE.html.
22. Quoted in Flanagan (2009).
23. *Real Time with Bill Maher* (March 21, 2008). Ironically, the panelist who suggested "moving on" was Jon Hamm who, at the time, played a serial womanizer on TV's *Mad Men*.
24. For more on Morgan's life and thought, see Moses (2008).
25. Morgan (1877/1908), p. 418, 427.
26. Darwin (1871/2007), p. 360.
27. Morgan (1877/1908), p. 52.
28. Dixson (1998), p. 37.

Chapter 3: A Closer Look at the Standard Narrative of Human Sexual Evolution

1. With apologies to John Perry Barlow, author of "A Ladies' Man and Shameless." At: http://www.nerve.com/personalEssays/Barlow/shameless/index.asp?page=1.
2. Wilson (1978), p. 148
3. Pinker (2002), p. 252.
4. Barkow et al. (1992), p. 289.
5. Barkow et al. (1992), pp. 267–268.
6. Acton (1857/62), p. 162.
7. Symons (1979), p. vi.
8. Bateman (1948), p. 365.
9. Clark and Hatfield (1989).
10. Wright (1994), p. 298.
11. Buss (2000), p. 140.
12. Wright (1994), p. 57.
13. Birkhead (2000), p. 33.
14. Wright (1994), p. 63.
15. Henry Kissinger—just our opinion. Nothing personal.
16. Wright (1994), pp. 57–58.
17. Symons (1979), p. v.
18. Fisher (1992), p. 187.

Chapter 4: The Ape in the Mirror

1. See Caswell et al. (2008) and Won and Hey (2004). Rapid advances in genetic testing have reopened the debate over the timing of the chimp/bonobo split. We use the widely accepted estimate of 3 million years, though it may turn out to have occurred less than a million years ago.
2. This account from de Waal and Lanting (1998).
3. Harris (1989), p. 181.
4. Symons (1979), p. 108.
5. Wrangham and Peterson (1996), p. 63.
6. Sapolsky (2001), p. 174.
7. Table based on de Waal (2005a) and Dixson (1998).
8. Stanford (2001), p. 116.
9. Berman (2000), pp. 66–67.
10. Dawkins (1976), p. 3.

11. http://www.edge.org/3rd_culture/woods_hare09/woods_hare09_index.html.
12. de Waal (2005), p. 106.
13. Theroux (1989), p. 195.
14. Pusey (2001), p. 20.
15. Stanford (2001), p. 26.
16. McGrew and Feistner (1992), p. 232.
17. de Waal (1995).
18. de Waal and Lanting (1998), p. 73.
19. de Waal (2001a), p. 140.
20. The quote appears here: http://primatediaries.blogspot.com/2009/03/bonobos-in-garden-of-eden.html.
21. Fisher (1992), p. 129.
22. Fisher (1992), pp. 129–130.
23. Fisher (1992). These quotes are all taken from an endnote on page 329.
24. Fisher (1992), p. 92.
25. Fisher (1992), pp. 130–131.
26. de Waal (2001b), p. 47.
27. de Waal (2005), pp. 124–125.
28. A true man of science, de Waal was kind enough to review and critique parts of this book, including sections where we disagree with some of his views.
29. The information in this chart is taken from various sources (Blount, 1990; Kano, 1980 and 1992; de Waal and Lanting, 1998; Savage-Rumbaugh and Wilkerson, 1978; de Waal, 2001a; de Waal, 2001b).

Part II: Lust in Paradise (Solitary)

Chapter 5: Who Lost What in Paradise?

1. For readers interested in further understanding how and why the shift from foraging to cultivation happened, Fagan (2004) and Quinn (1995) are both great places to start.
2. Cochran and Harpending (2009) point out some of these parallels: "In both [domesticated] humans and domesticated animals," they write, "we see a reduction in brain size, broader skulls, changes in hair color or coat color, and smaller teeth." (p. 112.)
3. Anderson is quoted in "Hellhole," by Atul Gawande in *The New Yorker*, March 30, 2009. The article is very much worth reading for its

examination of whether solitary confinement is so anti-human that it qualifies as torture. Gawande concludes it clearly does, writing, "Simply to exist as a normal human being requires interaction with other people."

4. Jones et al. (1992), p. 123.
5. Although only humans and bonobos appear to have sex throughout the menstrual cycle, both chimps and some types of dolphins seem to share our predilection for engaging in sex for pleasure, as opposed to reproduction alone.
6. These tidbits come from Ventura's wonderful essay on the origins of jazz and rock music, "Hear That Long Snake Moan," published in Ventura (1986). The book is out of print, but you can access this essay and other writing at Ventura's website: http://www.michaelventura.org/. The Thompson material can be found both in Ventura's essay and in Thompson (1984).

Chapter 6: Who's Your Daddies?

1. Harris (1989), p. 195.
2. Beckerman and Valentine (2002), p. 10.
3. Beckerman and Valentine (2002), p. 6.
4. Kim Hill is quoted in Hrdy (1999b), pp. 246–247.
5. Among the Bari people of Colombia and Venezuela, for example, researchers found that 80 percent of the children with two or more socially recognized fathers survived to adulthood, whereas only 64 percent of those with one official father made it that far. Hill and Hurtado (1996) reported that among their sample of 227 Aché children, 70 percent of those with only one recognized father survived to age ten, while 85 percent of those with both a primary and secondary father made it that far.
6. The quote is from an article by Sally Lehrman posted on AlterNet.org. Available at http://www.alternet.org/story/13648/?page=entire.
7. Morris (1981), pp. 154–156.
8. In Beckerman and Valentine (2002), p. 128.
9. See Erikson's chapter in Beckerman and Valentine (2002).
10. Williams (1988), p. 114.
11. Caesar (2008), p. 121.
12. Quoted in Sturma (2002), p. 17.

13. See Littlewood (2003).
14. At this point, naysayers will point out that Margaret Mead's famous claims of South Seas libertines were debunked by Derek Freeman (1983). But Freeman's debunking has been debunked as well, thus leaving Mead's original claims, what, *rebunked*? Hiram Caton (1990) and others have argued, quite compellingly, that Freeman's relentless attacks on Mead were likely motivated by a psychiatric disorder that also led to several paranoid outbursts of such intensity that he was forcibly removed from Sarawak by Australian diplomatic officials. The general consensus in the anthropological community seems to be that it's unclear to what extent, if any, Mead's findings were mistaken. Freeman's purported debunking took place after decades of Christian indoctrination of Samoans, so it should surprise no one if the stories he heard differed significantly from those told to Mead half a century earlier. For a brief review, we recommend Monaghan (2006).
15. Ford and Beach (1952), p. 118.
16. Small (1993), p. 153.
17. de Waal (2005), p. 101.
18. Morris (1967), p. 79.
19. http://primatediaries.blogspot.com/2007/08/forbidden-love.html.
20. Kinsey (1953), p. 415.
21. Sulloway (1998).
22. For a review of other mammals that practice sharing behavior, see Ridley (1996) and Stanford (2001).
23. Bogucki (1999), p. 124.
24. Knight (1995), p. 210.
25. The extent to which ovulation truly is hidden in humans is not as settled a matter as many authorities claim. There is good reason to believe that olfactory systems are still able to detect ovulation in women and that such systems are significantly atrophied when compared with those of ancestral humans. See, for example, Singh and Bronstad (2001). Furthermore, there is reason to believe that women advertise their fertility status via visual cues such as jewelry and changes in facial attractiveness. See, for example, Roberts et al. (2004).
26. Daniels (1983), p. 69.
27. Gregor (1985), p. 37.
28. Crocker and Crocker (2003), pp. 125–126.
29. Wilson (1978), p. 144.

Chapter 7: Mommies Dearest

1. Pollock (2002), pp. 53–54.
2. The quote is taken from an interview by Sarah van Gelder, "Remembering Our Purpose: An Interview with Malidoma Somé," *In Context: A Quarterly of Humane Sustainable Culture*, vol. 34, p. 30 (1993). Available online at http://www.context.org/ICLIB/IC34/Some.htm.
3. Hrdy (1999), p. 498.
4. Darwin (1871), p. 610.
5. Leacock (1981), p. 50.
6. http://www.slate.com/id/2204451/.
7. Erikson (2002), p. 131.
8. Chernela (2002), p. 163.
9. Lea (2002), p. 113.
10. Chernela (2002), p. 173.
11. Morris (1998), p. 262.
12. Malinowski (1962), pp. 156–157.
13. See Sapolsky (2005).
14. Drucker (2004).
15. Even Jean-Jacques Rousseau, poster-boy for the Romantic ideal of the Noble Savage, made use of these baby disposals. In 1785, Benjamin Franklin visited the hospital where Rousseau had deposited his five illegitimate children and discovered a mortality rate of 85 percent among the babies there ("Baby Food," by Jill Lepore, in *The New Yorker*, January 19, 2009).
16. McElvaine (2001), p. 45.
17. Betzig (1989), p. 654.

Chapter 8: Making a Mess of Marriage, Mating, and Monogamy

1. As we write this, Tiger Woods is being accused of having "slept with" more than a dozen women in cars, parking lots, on sofas. . . . Are we to think he's a narcoleptic?
2. de Waal (2005), p. 108.
3. Trivers's paper is seen as the foundational text in establishing the importance of male provisioning (investment) as a crucial factor in female sexual selection, among other things. It's well worth a read if you want a deeper understanding of the overall development of evolutionary psychology.
4. Ghiglieri (1999), p. 150.

5. Small (1993), p. 135.
6. Roughgarden (2007). Available online: http://www.redorbit.com/news/science/931165/challenging_darwins_theory_of_sexual_selection/index.html.
7. *The New Yorker*, November 25, 2002.
8. Cartwright's article is available here: http://www.pbs.org/wgbh/aia/part4/4h3106t.html.
9. Symons (1979), p. 108.
10. Valentine (2002), p. 188.
11. Article by Souhail Karam, *Reuters,* July 24, 2006.
12. *The New Yorker*, April 17, 2007.
13. Vincent of Beauvais *Speculum doctrinale* 10.45.
14. Both from Townsend and Levy (1990b).

Chapter 9: Paternity Certainty: The Crumbling Cornerstone of the Standard Narrative

1. Edgerton (1992), p. 182.
2. In Margolis (2004), p. 175.
3. Pollock (2002), p. 53.
4. For more on the deep connections between a society's levels of violence and its eroticism, see Prescott (1975).
5. Quoted in Hua (2001), p. 23.
6. Namu (2004), p 276. For an excellent look at Mosuo culture, check out *PBS Frontline World*, "The Women's Kingdom," available at www.pbs.org/frontlineworld/rough/2005/07/introduction_to.html.
7. Namu (2004), p. 69.
8. Namu (2004), p. 8.
9. This sacred regard for each individual's autonomy is characteristic of foragers, too. For example, when Michael Finkel visited the Hadza recently in Tanzania, he reported, "the Hadza recognize no official leaders. Camps are traditionally named after a senior male . . . but this honor does not confer any particular power. Individual autonomy is the hallmark of the Hadza. No Hadza adult has authority over any other." (*National Geographic*, December 2009.)
10. Hua (2001), pp. 202–203.
11. Namu (2004), pp. 94–95.

12. China's Kingdom of Women, Cynthia Barnes. Slate.com (November 17, 2006): http://www.slate.com/id/2153586/entry/2153614.
13. Goldberg (1993), p. 15.
14. (Photo: Christopher Ryan.) When I saw this old woman, I knew her face contained the feminine strength and humor I was hoping to convey in a photo. I gestured to ask if it would be all right to take her picture. She agreed, but asked me to wait, and immediately started calling. These two little girls (granddaughters? Great-granddaughters?) came running. Once she had them in her arms, she gave me the go-ahead to take the shot.
15. The book was published in 2002, while Goldberg's came out almost a decade earlier, but *all* of Sanday's work on the Minangkabau, including the paper Goldberg cites, argues *against* his position—a point certainly deserving of mention.
16. Source: http://www.eurekalert.org/pub_releases/2002-05/uop-imm050902.php.
17. Source: www.eurekalert.org/pub_releases/2002-05/uop-imm050902.php.
18. Most of these quotes are from an article by David Smith that appeared in *The Guardian*, September 18, 2005, available online at http://www.guardian.co.uk/uk/2005/sep/18/usa.filmnews, or Stephen Holden's review in *The New York Times*, June 24, 2005, available online at http://movies.nytimes.com/2005/06/24/movies/24peng.html?_r=2.
19. *The San Diego Union-Tribune*: "Studies Suggest Monogamy Isn't for the Birds—or Most Creatures," by Scott LaFee, September 4, 2002.
20. "Monogamy and the Prairie Vole," *Scientific American* online issue, February 2005, pp. 22–27.
21. Things have become a bit more muddled since Insel said that. More recently, Insel and others have been working on trying to discover the hormonal correlations underlying the fidelity or lack thereof among prairie, montane, and meadow voles. As reported in the October 7, 1993 issue of *Nature*, Insel and his team found that vasopressin, a hormone released during mating, seemed to trigger protective, nest-guarding behavior in some species of male voles, but not others, leading to speculation about "monogamy genes." See http://findarticles.com/p/articles/mi_m1200/is_n22_v144/ai_14642472 for a review. In 2008, Hasse Walum of the Karolinska Institute in Sweden found that a variation in the gene called *RS3 334* seemed to be associated

with how easily men bonded emotionally with their partners. Most interestingly, the gene appears to have some association with autism as well. The reference for Walum's paper is *Proceedings of the National Academy of Sciences*, DOI: 10.1073pnas.0803081105. A news article summarizing the findings is online at http://www.newscientist.com/article/dn14641-monogamy-gene-found-in-people.html.

Chapter 10: Jealousy: A Beginner's Guide to Coveting Thy Neighbor's Spouse

1. Darwin (1871/2007), p. 184.
2. Hrdy (1999b), p. 249.
3. Known to historians as The Wicked Bible or The Adulterous Bible, the mistake led to the royal printers losing their license and a £300 fine.
4. Confusingly, the tribe that came to be known as the Flatheads was not one of them, as their heads were "flat," like the white trappers', while the neighboring tribes' heads were bizarrely conical.
5. Grayscale reproduction scanned from Eaton, D.; Urbanek, S.: Paul Kane's Great Nor-West, University of British Columbia Press; Vancouver, 1995.
6. In fact, Maryanne Fisher and her colleagues found the opposite; distress was greater if the infidelity involved someone with familial bonds (see Fisher, et al. [2009]).
7. Buss (2000), p. 33.
8. Buss (2000), p. 58.
9. Jethá and Falcato (1991).
10. Harris (2000), p. 1084.
11. For an overview of Buss's research on jealousy, see Buss (2000). For research and commentary rebutting his work, see Ryan and Jethá (2005), Harris and Christenfeld (1996), and DeSteno and Salovey (1996).
12. www.epjournal.net/filestore/ep06667675.pdf.
13. Holmberg (1969), p. 161.
14. From an "On Faith" blog post in *The Washington Post*, November 29, 2007: http://newsweek.washingtonpost.com/onfaith/panelists/richard_dawkins/2007/11/banishing_the_greeneyed_monste.html.
15. Wilson (1978), p. 142.

Part III: The Way We Weren't

Chapter 11: "The Wealth of Nature" (Poor?)

1. Presumably, he was reading the sixth edition, published in 1826.
2. Barlow (1958), p. 120.
3. It's no accident that Darwin was well aware of Malthus's thinking. Harriet Martineau, an early feminist, economic philosopher, and outspoken opponent of slavery, had been close to Malthus before striking up a friendship with Darwin's older brother, Erasmus, who introduced her to Charles. Had Charles not been "astonished to find how ugly she is," some, including Matt Ridley, suspect their friendship might have led to marriage. It would surely have been a marriage with lasting effects on Western thought (see Ridley's article, "The Natural Order of Things," in *The Spectator*, January 7, 2009).
4. Shaw (1987), p. 53.
5. Darwin (1871/2007), p. 79. Both Malthus and Darwin would have profited from familiarity with MacArthur and Wilson's (1967) thoughts on r/K reproduction and selection. Briefly, they posit that some species (like many insects, rodents, etc.) reproduce quickly to fill an empty ecological niche. They don't expect most of their young to survive to adulthood, but they flood the environment quickly (r-selected). K-selected species have fewer young and invest heavily in each of them. Such species are generally in a state of Malthusian equilibrium, having already reached a population/food supply stasis point. Thus, these questions: as *Homo sapiens* is clearly a K-selected species, at what point did our environmental niche become saturated? Or have we continually found ways to expand our niche as human population expanded? If so, what does this mean for the underlying mechanisms of natural selection when applied to human evolution?
6. For example: "In the roughly 2 million years our ancestors lived as hunters and gatherers, the population rose from about 10,000 protohumans to about 4 million modern humans. If, as we believe, the growth pattern during this era was fairly steady, then the population must have doubled about every quarter million years, on average." *Economics of the Singularity,* Robin Hanson, http://www.spectrum.ieee.org/jun08/6274.
7. Source: U.S. Census Bureau: http://www.census.gov/ipc/www/worldhis.html.

8. Lilla (2007).
9. Smith's essay is online at http://realhumannature.com/?page_id=26.
10. Hassan (1980).
11. For a different take on how and why prehistoric population levels grew so slowly, see Harris (1977), particularly Chapter 2. For yet another take, see Hart and Sussman (2005), who argue that our ancestors did in fact live in Hobbesian fear—but not from each other so much as from constant predation. Malthus acknowledged the low population growth of Native Americans, but he attributed it to lack of libido caused by food shortages, "phlegmatic temperament," or "a natural defect in their bodily frame" (I. IV. P. 3).
12. Most of the other species of hominids that had spread from Africa to Asia and Europe previously were already long gone by the time modern humans wandered out of Africa. Those still hanging on, Neanderthals and (possibly) *Homo erectus*, would have been at a huge disadvantage if there was interspecies competition—which is unclear. One could argue that the presence of Neanderthals in Europe and parts of Central Asia may have led to competition over hunting areas, but the extent of contact between our ancestors and Neanderthals, if any, is unresolved. Also, any overlap would have been partial, as Neanderthals appear to have been top-level carnivores, while *Homo sapiens* are and were enthusiastic omnivores (see, for example, Richards and Trinkaus, 2009).
13. The question of when humans first arrived in the Americas is unresolved. Recent archaeological finds in Chile suggesting human settlements dating to about 35,000 years ago have thrown open the question of how and when the first humans arrived in the western hemisphere. See, for example, Dillehay et al. (2008).
14. See Amos and Hoffman (2009), for example. Paleoanthropologist John Hawks isn't convinced that population bottlenecks necessarily imply sparse prehistoric human populations overall, proposing that "many small groups of humans were in fact competing intensively, and many of them failed to persist over the long term. In other words, a small effective size is hardly evidence of no ancient competition or warfare. It may be the result of intense competition leading to many local extinctions" (see his blog: http://johnhawks.net/node/1894). Given the persistence of hunter-gather populations in the world's least habitable zones, the relative abundance of the rest of the planet, and

the genetic evidence of just a few hundred breeding pairs after the Toba eruption 70,000 years ago (Ambrose, 1998), we aren't convinced by Hawks's scenario of "many local extinctions" due to competition—as opposed to planetary catastrophe.

15. Agriculture itself can be seen as a response to ecological saturation brought about by the combined effects of gradually rising regional population and catastrophic climate change. For example, Nick Brooks, a researcher at the University of East Anglia, argues, "Civilization was in large part an accidental by-product of unplanned adaptation to catastrophic climate change." Brooks and others argue that the agricultural shift was a "last resort" response to deteriorating environmental conditions. For a comprehensive discussion of how climate change might have provoked agriculture, see Fagan (2004).
16. Known as "geophagy," dirt eating is common in societies around the world—especially among pregnant and lactating women. Additionally, many otherwise toxic foods containing poisonous alkaloids and tannic acids are cooked along with alkaloid-binding clays. Clay can be a rich source of iron, copper, magnesium, and calcium—all critical during pregnancy.
17. August 5, 2007.
18. http://moneyfeatures.blogs.money.cnn.com/2009/04/30/millionaires-arent-sleeping-well-either/?section=money_topstories.
19. See Wolf et al. (1989) and Bruhn and Wolf (1979). Malcolm Gladwell (2008) also discusses Roseto.
20. Sahlins (1972), p. 37.
21. http://www.newyorker.com/online/blogs/books/2009/04/the-exchange-david-plotz.html.
22. Malthus (1798), Book I, Chapter IV, paragraph 38.
23. Darwin (1871/2007), p. 208.
24. For a more detailed analysis of how modern economic theory plays out (or doesn't) among non-state societies, see Henrich et al. (2005) and Richard Lee's chapter titled "Reflections on Primitive Communism," in Ingold et al. (1988).

Chapter 12: The Selfish Meme (Nasty?)

1. In *A Theory of Moral Sentiments*, Smith wrote, "How selfish soever man may be supposed, there are evidently some principles in his

nature, which interest him in the fortune of others, and render their happiness necessary to him, though he derives nothing from it, except the pleasure of seeing it."

2. Gowdy (1998), p. xxiv.
3. From Mill (1874).
4. *New York Times*, July 23, 2002, "Why We're So Nice: We're Wired to Cooperate." http://www.nytimes.com/2002/07/23/science/why-we-re-so-nice-we-re-wired-to-cooperate.html. For the original research, see Rilling et al. (2002).
5. We've drawn from an excellent analysis of Hardin's paper by Ian Angus, which can be found at http://links.org.au/node/595.
6. See Ostrom (2009), for example.
7. See Dunbar (1992 and 1993).
8. Harris (1989), pp. 344–345.
9. Bodley (2002), p. 54.
10. Harris (1989), p. 147.
11. van der Merwe (1992), p. 372. Also see Jared Diamond, "The Worst Mistake in the History of the Human Race" (widely available online; see here, for example: http://www.awok.org/worst-mistake/).
12. Le Jeune (1897), pp. 281–283.
13. Gowdy (1998), p. 130.
14. Quoted in Menzel and D'Aluisio, p. 178.
15. Harris (1977), p. x. Also see Eaton, Shostak, and Konner (1988).
16. Gowdy (1998), p. 13.
17. Gowdy (1998), p. 23.
18. Harris (1980), p. 81.
19. Ridley (1996), p. 249.
20. See de Waal (2009) for much more on the biological origins of empathy and instinctive justice.
21. Dawkins (1998), p. 212.
22. de Waal and Johanowicz (1993).
23. Sapolsky and Share (2004). Also see Natalie Angier, "No Time for Bullies: Baboons Retool Their Culture," *New York Times*, April 13, 2004.
24. Boehm (1999), p. 3, 68.
25. Fromm (1973), p. 60.
26. Gowdy (1998), p. xvii.

Chapter 13: The Never-Ending Battle over Prehistoric War (Brutish?)

1. From his closing argument in the Scopes case.
2. Wade (2006), p. 151.
3. Recent studies of mitochondrial DNA suggest that even before the human migrations out of Africa that began about 60,000 years ago, human populations were largely isolated from each other for as much as 100,000 years, localized in eastern and southern Africa. Only about 40,000 years ago did these two lines reunite, becoming a single pan-African population, according to this research. See Behar et al. (2008). Full paper available online at http://www.cell.com/AJHG/fulltext/S0002-9297%2808%2900255-3#.
4. Readers interested in further exploration of the critique of Hobbesian assumptions regarding war in prehistory could begin with Fry (2009) and Ferguson (2000).
5. Pinker's talk was based upon an argument he presents in *The Blank Slate* (2002), particularly in the last few pages of the third chapter.
6. The link to Pinker's presentation is http://www.ted.com/index.php/talks/steven_pinker_on_the_myth_of_violence.html. You can find many other interesting presentations at this site. You might want to search Sue Savage-Rumbaugh's talks on bonobos, for example. If you prefer to read Pinker's remarks, an essay based upon the talk can be found at www.edge.org/3rd_culture/pinker07/pinker07_index.html.
7. Note that Pinker's chart represents part of a chart in Keeley's book (1996), and that Keeley refers to these societies as "primitive," "prestate," and "prehistoric" in his charts (pp. 89–90). Indeed, Keeley distinguishes what he calls "sedentary hunter-gatherers" from true "nomadic hunter-gatherers," writing, "Low-density, nomadic hunter-gatherers, with their few (and portable) possessions, large territories, and few fixed resources or constructed facilities, had the option of fleeing conflict and raiding parties. At best, the only thing they would lose by such flight was their composure" (p. 31).

   As we've established, these nomadic (immediate-return) hunter-gatherers are most representative of human prehistory—a period that is, *by definition*, before the advent of settled communities, cultivated food, domesticated animals, and so on. Keeley's confusion (and thus, Pinker's) is largely due to his referring to horticulturalists, with their gardens, domesticated animals, and settled villages, as "sedentary hunter-gatherers." Yes, they *do* occasionally hunt and they sometimes gather, but because these activities are not their *sole* source of food, their lives are dissimilar to those of immediate-return hunter-gatherers. Their gardens, settled villages, and so on make territorial defense necessary and fleeing conflict much more problematic than it was for our ancestors. They—unlike true immediate-return foragers—have a lot to lose by simply fleeing aggression.

Keeley acknowledges this crucial difference, writing, "Farmers and sedentary hunter-gatherers had little alternative but to meet force with force or, after injury, to discourage further depredations by taking revenge" (p. 31).

The point bears repeating. If you live a settled life in a stable village, have a labor-expensive shelter, cultivated fields, domesticated animals, and too many possessions to carry easily, *you're not a hunter-gatherer.* Prehistoric human beings did not have any of these things, which is, after all, precisely what made them "prehistoric." Pinker either fails to appreciate this essential point, or simply ignores it.

8. Societies in Pinker's chart:

Jivaro	The Jivaro cultivate yams, peanuts, sweet manioc, maize, sweet potatoes, peanuts, tuber beans, pumpkins, plantains, tobacco, cotton, banana, sugarcane, taro, and yam. They also traditionally domesticate llamas and guinea pigs and later the introduced dog, chicken, and pig.
Yanomami	The Yanomami are foraging, "slash-and-burn" horticulturists. They cultivate plantains, cassava, and bananas.
Mae Enga	The Mae Enga grow sweet potatoes, taro, bananas, sugarcane, Pandanus nuts, beans, and various leaf greens, as well as potatoes, maize, and peanuts. They raise pigs, used not only for meat but for important ritualistic celebrations.
Dugum Dani	About 90 percent of the Dani diet is sweet potatoes. They also cultivate banana and cassava. Domestic pigs are important both for currency used in barter and for the celebration of important events. Pig theft is a major cause of conflict.

Murngin	The Murngin economy was based primarily on fishing, collecting shellfish, hunting, and gathering until the establishment of missions and the gradual introduction of market goods in the 1930s and 1940s. While hunting and gathering remain important for some groups, motor vehicles, aluminum boats with outboard engines, guns, and other introduced tools have replaced indigenous techniques.
Huli	The Huli's staple food is the sweet potato. Like other groups in Papua New Guinea, the Huli prize domestic pigs for meat and status.

9. This, according to Fry (2009).
10. Knauft (1987 and 2009).
11. To make matters even worse, Pinker juxtaposes these bogus "hunter-gatherer" mortality rates with a tiny bar showing the relatively few war-related deaths of males in twentieth-century United States and Europe. This is misleading in many respects. Perhaps most important, the twentieth century gave birth to "total war" between nations, in which civilians (not just male combatants) were targeted for psychological advantage (Dresden, Hiroshima, Nagasaki . . .), so counting only male mortalities is meaningless.

    Furthermore, why did Pinker not include the tens of millions who died in some of the most vicious and deadly examples of twentieth-century warfare? In his discussion of "our most peaceful age," he makes no mention of the Rape of Nanking, the entire Pacific theater of World War II (including the detonation of two nuclear bombs over Japan), the Khmer Rouge and Pol Pot's killing fields in Cambodia, several consecutive decades-long wars in Vietnam (against the Japanese, French, and Americans), the Chinese revolution and civil war, the India/Pakistan separation and subsequent wars, or the Korean war. None of these many millions are included in his assessment of twentieth-century (male) war fatalities.

    Nor does Pinker include Africa, with its never-ending conflicts, child soldiers, and casual genocides. No mention of Rwanda. Not a Tutsi or Hutu to be found. He leaves out every one of South America's various twentieth-century wars and dictatorships infamous for torturing and disappearing tens of thousands of civilians. El Salvador?

Nicaragua? More than 100,000 dead villagers in Guatemala? *Nada. Absolutamente nada.*

12. For example, see Zihlman et al. (1978 and 1984).
13. *Why War?* available online at http://realhumannature.com/?page_id=26. After we contacted him to ask how he could possibly justify the omission, Smith at first cited Wrangham and Peterson's dismissal of bonobos as being less representative than chimps of our last common ancestor. When we pointed out that many primatologists argue that bonobos are probably *more* representative, that even Wrangham had revised his opinion on the point, and that in any case, *it is factually wrong to say that chimps are our "closest non-human relative" without mentioning bonobos*, he finally relented and added two brief references to bonobos to his lurid descriptions of chimps' "bloody wars of attrition." Since the online essay was an extract from his book, which was already in print, it seems unlikely these reluctant changes are reflected there.
14. Ghiglieri (1999), pp. 104–105.
15. For a review, see Sussman and Garber's chapter in Chapman and Sussman (2004).
16. The quote is from de Waal (1998), p. 10.
17. Goodall (1971), cited in Power (1991), pp. 28–29.
18. Strangely, even though he agrees with this central point made by Power, de Waal barely mentions her work—and only to dismiss her, at that. In an endnote in his 1996 book, *Good Natured: The Origins of Right and Wrong in Humans and Other Animals*, he writes, "On the basis of her reading of the literature, Power (1991) has argued that provisioning at some field sites (such as Gombe's banana camp) turned the chimpanzees more violent and less egalitarian, and thus changed the 'tone' of relationships both within and between communities. Power's analysis—which blends a serious reexamination of available data with nostalgia for the 1960s image of apes as noble savages—raises questions that will no doubt be settled by ongoing research on unprovisioned wild chimpanzees."

    This dismissal of Power's analysis strikes us as unjustified and uncharacteristically ungenerous. Regardless of whether or not she felt "nostalgia for the 1960s" (an emotion we didn't detect in her book), de Waal admits her analysis "raises questions" that merit investigation. These questions threaten to recast a great deal of data concerning chimpanzee social interactions—of great interest to de Waal, one of the world's leading authorities on chimpanzee behavior and a man whose scholarship demonstrates deep respect for critical analysis.

19. Ghiglieri (1999), p. 173.
20. For a review of these reports and a rebuttal to Power's argument, see Wilson and Wrangham (2003). The paper is available online at http://anthro.annualreviews.org.
21. Nolan (2003).
22. Behar et al. (2008). Also, for an excellent review of this material, see Fagan (2004).
23. Turchin (2003 and 2006).
24. Readers with mental images of Sioux (Lakota) chiefs with eagle-feather war bonnets rippling in the wind should keep in mind that in the generations before first contact with whites, disease spread through many tribes and the arrival of horses brought severe cultural disruptions, leading to conflict between groups that had been at peace previously (see Brown, 1970/2001).
25. Edgerton (1992), pp. 90–104.
26. Ferguson (2003).
27. On Christmas Day, 1968, Apollo 8 astronaut Frank Borman read this prayer to a world audience: "Give us, O God, the vision which can see thy love in the world in spite of human failure. Give us the faith to trust the goodness in spite of our ignorance and weakness. Give us the knowledge that we may continue to pray with understanding hearts, and show us what each one of us can do to set forth the coming of the day of universal peace. Amen."
28. Tierney (2000), p. 18. Tierney's book sparked a conflagration that makes any chimpanzee community seem downright pacific in comparison. The bulk of the controversy concerns Tierney's charges that Chagnon and his colleague, James Neel, may have caused a fatal epidemic among the Yanomami. Not having examined these charges in detail, we have nothing to add to that discussion, limiting our critique to Chagnon's methodology and scholarship as it applies to Yanomami warfare.
29. By comparison, Chagnon's total time among the Yanomami adds up to about five years. Readers interested in knowing more about the Yanomani might start with Good (1991). This is a very personal and accessible account of his time living with them (and ultimately, taking a wife there). Tierney (2000) outlines the case against Chagnon, though going far beyond the critiques we've outlined here. Ferguson (1995) offers an in-depth analysis of Chagnon's calculations and conclusions. For more of Ferguson's views on the origins of war, the following two papers can be downloaded from his departmental web page (http://

andromeda.rutgers.edu/socant/brian.htm): *Tribal, "Ethnic," and Global Wars*, and *Ten Points on War*, which includes a broad discussion of biology, archaeology, and the Yanomami controversy. Borofsky (2005) offers a balanced account of the controversy and the context in which it occurred. Of course, Chagnon's work is readily available as well.

30. Quoted in Tierney (2000), p. 32.
31. *Washington Post* review of *Darkness in El Dorado: Jungle Fever*, by Marshall Sahlins, Sunday, December 10, 2000, p. X01.
32. Chagnon (1968), p. 12.
33. Tierney (2000), p. 14.
34. Sponsel (1998), p. 104.
35. October 23, 2008.

Chapter 14: The Longevity Lie (Short?)

1. Note that these numbers are for demonstration purposes only. To keep it simple (and since it's meaningless anyway), we haven't adjusted for male/female size differences, regional variations in average infant skeleton sizes, and so on.
2. October 6, 2008.
3. Adovasio et al. (2007), p. 129.
4. Gina Kolata, "Could We Live Forever?," November 11, 2003.
5. *Scientific American*, March 6, p. 57.
6. Harris (1989), pp. 211–212.
7. http://www.gendercide.org/case_infanticide.html.
8. These numbers don't include the selective abortion of female fetuses, which is widespread in these countries. For example, Agence-France Presse reports that selective abortions have left China with 32 million more men than women, and that in just one year (2005), more than 1.1 million more boys than girls were born in China.
9. Philosopher Peter Singer has written thought-provoking books and essays on the question of how to calculate the value of human versus nonhuman life. See, for example, Singer (1990).
10. Cited in Blurton Jones et al. (2002).
11. Blurton Jones et al. (2002).
12. See Blurton Jones et al. (2002).
13. An excellent paper highly recommended to readers interested in these matters is Kaplan et al. (2000). The paper can be downloaded at Kaplan's faculty website: www.unm.edu/~hebs/pubs_kaplan.html.
14. From the paper by Kaplan et al. cited above, p. 171.

15. Readers interested in seeing how these same agricultural curses are playing out in the modern world might want to read Michael Pollan's *In Defense of Food: An Eater's Manifesto* (2009).
16. Larrick et al. (1979).
17. Source: Diamond (1997).
18. Edgerton (1992), p. 111.
19. Cohen et al. (2009).
20. Horne et al. (2008).
21. While we're on the subject of hammocks, we'd like to take this opportunity to formally propose that hammocks—not spear points or stone blades—were the first example of human technology. That no hard evidence for this proposal has been unearthed is due to hammocks being made of perishable fibers. (Who wants a stone hammock?) Even chimps and bonobos fashion primitive hammocks by weaving together tree branches for sleeping platforms.
22. See Sapolsky (1998) for an excellent overview of how stress affects us. On the question of human/bonobo similarities concerning stress, it's interesting to note that when bombs fell near them in World War II, *all* the bonobos in the zoo died from the stress the explosions caused, while *none* of the chimps perished (according to de Waal and Lanting, 1998).
23. *The New Yorker*, June 26, 2006, p. 76.

Part IV: Bodies in Motion

1. This quote is taken from a debate between Gould on one side and Steven Pinker and Daniel Dennett on the other. Well worth reading, if you like high-brow discussion with plenty of low blows, is "Evolution: The Pleasures of Pluralism," *The New York Review of Books* 44(11): 47–52.
2. Potts (1992), p. 327.

Chapter 15: Little Big Man

1. Miller (2000), p. 169.
2. Though not always, as other factors can influence body-size dimorphism, apart from the intensity of male-male mating conflict. See Lawler (2009), for example.
3. Male *Australopithecus* (between three and four million years ago) are thought to have been about fifty percent larger than females. Recent

papers suggest *Ardipethicus ramidus*, another supposed human ancestor (thought to be a million or so years older than *Australopithecus*) was closer to our 15 to 20 percent levels. But keep in mind that the much-ballyhooed reconstruction of *Ardipethicus ramidus* relied upon bits and pieces of many different individuals, so our sense of body size dimorphism 4.4 million years ago is based upon educated guesses, at best (White et al., 2009).

4. Lovejoy (2009).
5. http://www.psychologytoday.com/articles/200706/ten-politically-incorrect-truths-about-human-nature.
6. Supplemental note. On sexual selection in relation to monkeys. Reprinted from *Nature*, November 2, 1876, p. 18. http://sacred-texts.com/aor/darwin/descent/dom25.htm.
7. As we'll discuss in the next chapter, the *genital echo theory* posits that women developed pendulous breasts so that the cleavage would mimic the (is there a scientific term for this?) butt-crack that so enticed our primate ancestors. Following that line of reasoning, some argue that fancily named lipstick serves to re-create the bright red "hinder ends" that so perplexed poor Darwin.
8. See Baker and Bellis (1995) or Baker (1996) for the sperm-team theory.
9. Hrdy (1996) is a wonderfully erudite and engaging discussion of how some of Darwin's personal sexual hang-ups are reflected still in evolutionary theory.
10. Supplemental note. On sexual selection in relation to monkeys. Reprinted from *Nature*, November 2, 1876, p. 18. http://sacred-texts.com/aor/darwin/descent/dom25.htm.
11. Diamond (1991), p. 62.

Chapter 16: The Truest Measure of a Man

1. de Waal (2005), p. 113.
2. In Barkow et al. (1992), p. 299.
3. Barash and Lipton (2001), p. 141.
4. Pochron and Wright (2002).
5. Wyckoff et al. (2000). Other research looking into primate testicular genetics has reinforced the impression that ancestral human mating behavior more closely resembled the promiscuity of chimps than the one-male-at-a-time gorillas. See, for example, Kingan et al. (2003), who conclude that although "predicting the expected intensity of

sperm competition in ancestral *Homo* is controversial, . . . we find patterns of nucleotide variability at SgI in humans to resemble more closely the patterns seen in chimps than in gorillas."

6. Short (1979).
7. Margulis and Sagan (1991), p. 51.
8. Lindholmer (1973).
9. For more on this, see work by Todd Shackelford, particularly Shackelford et al. (2007). Shackelford generously makes most of his published work available for free download at http://www.toddkshackelford.com/publications/index.html.
10. Symons (1979), p. 92. Although we probably disagree with half of his conclusions, and much of the science is out of date, Symons's book is well worth reading for its wit and artistry alone.
11. Harris (1989), p. 261.
12. Sperm competition is an area of passionate debate. Space limitations (and quite possibly, readers' interest) preclude us from a more thorough discussion—especially concerning the highly controversial claims of Baker and Bellis regarding sperm teams composed of specialized cells acting as "blockers," "kamakaze," and "egg-getters." For a scientific review of their findings, see Baker and Bellis (1995). For a popularized review, see Baker (1996). For a balanced discussion of the controversy written by a third party, see Birkhead (2000), especially pp. 21–29.
13. Data primarily from Dixson (1998).
14. See, for example, Pound (2002).
15. Kilgallon and Simmons (2005).
16. Some readers will argue that these conventions in contemporary pornography are expressions of female subjugation and degradation rather than eroticism. Whether or not this is the case (a discussion we're going to sidestep at this juncture), one must still ask why it is being expressed in this way, with these images, given that there are so many ways of visibly humiliating a person. Some authorities believe the practice of *bukkake* originated as a way of punishing adulterous women in Japan—a somewhat less Puritanical *Scarlet Letter*, if you will (see, for example, "Bake a Cake? Exposing the Sexual Practice of Bukkake," poster presented at the 17th World Congress of Sexology, by Jeff Hudson and Nicholas Doong: http://abstracts.co.allenpress.com/pweb/sexo2005/document/50214). If you don't know what *bukkake* is,

and you're even slightly prone to being offended, please forget we even mentioned it.

Chapter 17: Sometimes a Penis Is Just a Penis

1. Frans de Waal suspects that bonobos have longer penises than humans, at least relative to body size, but most other primatologists seem to disagree with his assessment. In any case, there is no question that the human penis is far thicker than that of any other ape, in absolute terms or relative to body size, and far longer than that of any primate not clearly engaged in extreme sperm competition.
2. Sherfey (1972), p. 67.
3. One species of gibbon, the black-crested gibbon (*Hylobates concolor*), does in fact have an external, pendulous scrotum. Interestingly, this type of gibbon may also be exceptional in *not* being strictly monogamous (see Jiang et al., 1999).
4. Gallup (2009) offers an excellent summary of this material.
5. Dindyal (2004).
6. http://news.bbc.co.uk/go/pr/fr/-/2/hi/health/7633400.stm 2008/09/24.
7. Harvey and May (1989), p. 508.
8. Writing in the *Encyclopedia of Human Evolution*, Robert Martin notes, "Relative to body size, humans have a very low value for rmax—even in comparison with other primates. This suggests that selection has favoured a low breeding potential during human evolution. Any model of human evolution should take this into account." A low rmax value along with the very high levels of sexual activity typical of humans is yet another indication that sex has long functioned for nonreproductive purposes in our species.

   Similarly, while Dixson (1998) characterizes the seminal vesicles of monogamous and polygynous primates (except the gelada baboon) as vestigial or small, he classifies the human seminal vesicles as medium—noting that "it is reasonable to propose that natural selection might have favoured reduction in size of the vesicles under conditions where copulation is relatively infrequent and the need for large ejaculate volume and coagulum formation is reduced." He goes on to propose that "this might explain the very small size of the vesicles in primarily monogamous [primates]."
9. *BBC News* online, July 16, 2003.

10. *BBC News* online, October 15, 2007.
11. *Psychology Today*, March/April 2001.
12. Barratt et al. (2009).
13. Hypothetically, one could try to falsify this hypothesis using data on testicular volume and sperm production from some of the societies we've discussed where sperm competition and partible paternity are in effect. To this end, we've contacted every anthropologist we could locate who has worked in the Amazon (or anywhere else with hunter-gatherers), but no one seems to have managed to gather these delicate data. Still, even if it were found that males in these societies showed higher testicular volume and sperm production, as our hypothesis predicts, definitive confirmation of the hypothesis would be precluded by the relative absence of the environmental toxins that are presumably at least partly responsible for testicular atrophy in industrialized societies.
14. *BBC News* online, December 8, 2006.
15. Diamond (1986).
16. W. A. Schonfeld, "Primary and Secondary Sexual Characteristics. Study of Their Development in Males from Birth through Maturity, with Biometric Study of Penis and Testes," *American Journal of Diseases in Children* 65, 535–549 (cited in Short, 1979).
17. Harvey and May (1989).
18. Baker (1996), p. 316.
19. Bogucki (1999), p. 20.

Chapter 18: The Prehistory of O

1. Since this book was first published in 2010, Maines's research has been called into question. In 2018, the *Atlantic* published an article by Robinson Meyer and Ashley Fetters (*Victorian-Era Orgasms and the Crisis of Peer Review*) showing that Maines's contention that doctors treated hysteria with vibrators is unfounded and that her sourcing for this claim was sloppy at best. We've made minor changes to the text in light of the controversy.
2. Quotes taken from Margolis (2004).
3. See Money (2000). Interestingly, semen depletion is central to the ancient Taoist understanding of male health and sexuality as well. See, for example, Reid (1989).
4. On Baker Brown, see Fleming (1960) and Moscucci (1996).

5. Coventry (2000).
6. Although the clitoris is often referred to as "the only organ in the human body whose sole function is to provide pleasure," there are two problems with this observation. First, if female orgasm (pleasure) is functional in the senses we outline (increases chances of fertilization, inspires vocalizations, and thereby promotes sperm competition), then there is clearly a purpose to the pleasure. Secondly, what about male nipples? Not all men find them to be a site of pleasure, but they are certainly highly enervated and serve no functional purpose.
7. Margolis (2004), pp. 242–243.
8. Ironically, according to archaeologist Timothy Taylor (1996), this image of the Devil is thought to be derived from *Cernunnos*, the horned god, who was the Celtic translation of Indian tantric practice and thus originally a symbol of spiritual transcendence via sexual practice.
9. Coventry (2000).
10. Hrdy (1999b), p. 259.
11. Sherfey (1972), p. 113.

Chapter 19: When Girls Go Wild

1. Pinker (2002), p. 253.
2. Not to exclude women or gay men, but there is a dearth of scientific data on this particular angle. Interestingly, though, several people have reported to us anecdotally that when they've overheard their neighbors (both gay male and lesbian couples) having sex, the partner they considered to be the more feminine was the one who was making more noise.
3. When the director, Rob Reiner, showed the screenplay to his mother, she suggested that at the end of that scene, the camera cut to an older woman in the restaurant about to order, who says, "I'll have what she's having." The line was so brilliant that Reiner told his mother he'd insert it, but only if *she* agreed to deliver the line in the film, which she did.
4. Semple (2001).
5. Small (1993), p. 142.
6. Small (1993), p. 170.
7. Dixson (1998), pp. 128–129.
8. Pradhan et al. (2006).
9. These quotes are from Hamilton and Arrowood (1978).

10. The intensity of the female's vocalizations could, for example, guide the discerning male's orgasmic response—thus increasing the chances of simultaneous or near-simultaneous orgasm. As we discuss below, there is evidence such timing could be to the male's reproductive advantage.
11. The title, far from being the frat-boy declaration it may seem ("Without tits, there is no paradise."), is the name of a Colombian television drama about young women who get breast implants hoping to attract the attentions of local drug lords and thereby escape poverty.
12. For example, Symons (1979) and Wright (1994).
13. See Morris (1967), Diamond (1991), and Fisher (1992).
14. http://dir.salon.com/story/mwt/style/2002/05/28/booty_call/.
15. Though they can be considered permanently swollen, this is not to say that breasts don't change throughout a woman's life (and menstrual) cycle. They typically swell further at pregnancy, menstruation, and orgasm (up to 25 percent greater than normal, according to Sherfey), and diminish in size and fullness with age and breastfeeding.
16. Small (1993), p. 128.
17. Haselton et al. (2007). Available online at www.sciencedirect.com.
18. Many accounts of human sexuality incorporate this explanation, but that of Desmond Morris is probably still the most widely known.
19. Dixson (1998), pp. 133–134.
20. Dixson refers specifically to macaques and chimps in this passage, though he's speaking of the capacity for multiple orgasm in female primates in general in the section where the passage appears. Passages like this led us to wonder why Dixson hadn't followed the data to where they seem to so clearly lead. We sent him an email outlining our argument and requesting his comments and criticisms, but if he received our message, he chose not to respond.
21. Symons (1979), p. 89.
22. Lloyd, a former student of Stephen Jay Gould, recently published an entire book in which she reviews (and rather contemptuously dismisses) the various adaptive arguments for the female orgasm (*The Case of the Female Orgasm: Bias in the Science of Evolution*). For a sense of why we don't recommend her book, take a look at David Barash's review, available online ("Let a Thousand Orgasms Bloom"). Download at http://www.epjournal.net/filestore/ep03347354.pdf.
23. As noted above, some of the findings of Baker and Bellis are highly controversial. We mention them because they are known to many in

the general audience, but none of their findings are necessary to our argument.

24. Barratt et al. (2009). Available online at http://jbiol.com/content/8/7/63.
25. Pusey (2001).
26. Both quotes appear in Potts and Short (1999). The first quote is from the main text, page 38, and the second is quoting Laura Betzig, p. 39.
27. Dixson (1998), pp. 269–271. An excellent review of the development of the concept of postcopulatory sexual selection can be found in Birkhead (2000). Copious evidence for this filtering function can be found in Eberhard (1996), where the author presents dozens of examples of females exerting "post-copulatory control" over which sperm fertilize their eggs.
28. Dixson (1998), p. 2.
29. Small (1993), p. 122.
30. Gallup et al. (2002).

Part V: Men Are from Africa, Women Are from Africa

1. Wright (1994), p. 58.

Chapter 20: On Mona Lisa's Mind

1. Kendrick et al. (1998).
2. Baumeister (2000).
3. Chivers et al. (2007).
4. Much of the research reviewed here is mentioned in Bergner's excellent article "What Do Women Want?—Discovering What Ignites Female Desire," January 22, 2009. Link: http://www.nytimes.com/2009/01/25/magazine/25desire-t.html.
5. Anokhin et al. (2006).
6. Georgiadis et al. (2006). Or, for a review: Mark Henderson, "Women Fall into a 'Trance' During Orgasm," *Times Online*, June 20, 2005. http://www.timesonline.co.uk/tol/life_and_style/health/article535521.ece.
7. Tarin and Gómez-Piquer (2002).
8. Little's quote is from BBC News article: http://news.bbc.co.uk/2/hi/health/2677697.stm.
9. Wedekind et al. (1995). A more recent follow-up that confirms these results is Santos et al. (2005).

10. Birth control pills don't just interfere with women's ability to sense MHC in men, but appear to affect other feedback systems as well. See Laeng and Falkenberg (2007), for example.
11. For a recent survey of this research, see Alvergne and Lummaa (2009).
12. This isn't meant as an indictment of the pill. But in light of these changes, we'd strongly recommend that couples spend several months together using alternate forms of birth control before making long-term plans.
13. Lippa (2007). Available online at http://psych.fullerton.edu/rlippa/bbc_sexdrive.htm.
14. See Safron et al. (2007). A good review of related research is here: http://www.wired.com/medtech/health/news/2004/04/63115?currentPage=all.
15. Alexander and Fisher (2003).

Chapter 21: The Pervert's Lament

1. Dixson (1998), p. 145.
2. Both these interviews appear on NPR's *This American Life*, Episode #220. Available via free download at iTunes or at www.thislife.org.
3. According to Reid (1989), it was considered wise and healthful for young Chinese men to share their abundant sexual energies with older women, who would benefit from absorbing the energy released by male orgasm; likewise, it was felt that young women's orgasms would infuse older men with increased vitality. The same pattern is found in some foraging societies, as well as among some South Pacific island cultures.
4. One example among many: Dabbs et al. (1991, 1995) found, "Offenders high in testosterone committed more violent crimes, were judged more harshly by the parole board, and violated prison rules more often than those low in testosterone."
5. Gibson (1989).
6. One wonders about the long-term social repercussions of widespread sexual frustration in adolescent males. To what extent, for example, is this frustration a contributing factor to the misogynistic rage many men experience? How does this frustration affect young men's willingness to fight wars or join street gangs? While we don't agree with arguments like those advanced by Kanazawa (2007) claiming that Islam sanctions polygyny in order to increase the male sexual frustration that creates a pool of available suicide bombers, it's hard to dismiss the notion that intense frustration will often be expressed as misdirected rage.

7. Georgia has a serious problem with oral sex. Until 1998, it was illegal—even between a married couple in their own bedroom—and punishable by up to twenty years in prison.
8. For example, http://www.npr.org/templates/story/story.php?storyId=102386952&ft=1&f=1001.
9. Fortenberry (2005).
10. All quotes from this section are taken from Prescott (1975).
11. See Elwin (1968) and Schlegel (1995).
12. "Some Thoughts on the Science of Onanism," a speech delivered to the Stomach Club, a society of American writers and artists.
13. Money (1985).
14. See http://www.cirp.org/library/statistics/USA/.
15. Money (1985), pp. 101–102.
16. These men believed that any spices or strong flavors excited sexual energies, so they recommended bland diets to dampen the libido. Graham crackers and unsweetened breakfast cereal were originally marketed to parents of adolescent boys as foods that would help them evade the evils of masturbation. For a fictionalized—though largely accurate—depiction of these men and their movement, see Boyle (1993).
17. Interestingly, Freud's nephew, Edward Bernays, is considered one of the founders of public relations and modern advertising. Among his many famous ad campaigns was the first to associate cigarettes with increased autonomy for women. In the 1920s, Bernays staged a legendary publicity stunt still taught in business classes today. He arranged to have fashion models march in New York's Easter parade, each with a lit cigarette and wearing a banner calling it a "torch of liberty." For more on this, see Ewen (1976/2001).
18. Farmers know that in order to get a bull to mate with the same cow more than a few times, the bull has to be tricked into thinking it's a different cow. They do this by rubbing a blanket on another cow to absorb her scent and then throwing it on top of the cow to be mated. If the bull isn't fooled into it, he'll simply refuse—no matter how attractive the cow may be.
19. Sprague and Quadagno (1989).
20. See, for example, the documentary film *Rent a Rasta*, written and directed by J. Michael Seyfert: www.rentarasta.com, or the feature film *Heading South*, directed by Laurent Cantet, about women going to Haiti in the 1970s.

21. *The New Yorker*, July 6 and 13, 2009, p. 68.
22. Additionally, the so-called *Westermarck effect* appears to strongly dissuade sex between close familiars.
23. See, for example, Gray et al. (1997 and 2002) and Ellison et al. (2009).
24. See, for example, Glass and Wright (1985).
25. Roney et al. (2009), but also see Roney et al. (2003, 2006, and 2007).
26. Davenport (1965).
27. Kinsey et al. (1948), p. 589.
28. Symons (1979), p. 232.
29. Bernard (1972/1982).
30. Berkowitz and Yager-Berkowitz (2008).
31. Symons (1979), p. 250.
32. See, for example, Roney et al. (2003). Regular aerobic exercise, lots of garlic, stress avoidance, and plenty of sleep are also good ways to "keep it up." We should note that despite the anecdotal evidence, few scientists have risked ridicule by applying for grants to study the hormonal changes in philanderers. The phenomenon is well documented in other mammals, however (see, for example, Macrides et al., 1975). It's possible the effect may be mediated not by actual intercourse so much as by pheromones, which might explain the *bulusela* shops where Japanese men purchase girls' vaccum-packed (but used) panties from vending machines. Enterprising graduate students might want to consider research similar to Wedekind's "Sweaty T-shirt study," but with women's panties instead of men's shirts in the plastic bags, to see whether exposure to novel women's genital pheromones alone is enough to affect testosterone blood concentrations in men.
33. For example, for depression: Shores et al. (2004); heart disease: Malkin et al. (2003); dementia: Henderson and Hogervorst (2004); mortality: Shores et al. (2006).
34. Squire quoted by Phillip Weiss in his provocative article in *New York* magazine: "The affairs of men: The trouble with sex and marriage." May 18, 2008. Available here: http://nymag.com/relationships/sex/47055.

Chapter 22: Confronting the Sky Together

1. Wilson (1978), p. 148.
2. Holmberg (1969), p. 258.

3. "Is There Hope for the American Marriage?" Caitlin Flanagan *Time*, July 2, 2009. http://www.time.com/time/nation/article/0,8599,1908243,00.html.
4. These quotes from Druckerman were taken from a review of her book in *The Observer*, July 8, 2007.
5. Jaynes (1990), p. 67.
6. "What does marriage mean?" Dan Savage. In Salon.com, July 17, 2004: http://www.salon.com/mwt/feature/2004/07/17/gay_marriage/index.html.
7. Squire quoted by Weiss in *New York* magazine: "The affairs of men: The trouble with sex and marriage." May 18, 2008. Available here: http://nymag.com/relationships/sex/47055.
8. "Only You. And You. And You. Polyamory—relationships with multiple, mutually consenting partners—has a coming-out party." By Jessica Bennett. *Newsweek* (Web Exclusive) July 29, 2009. http://www.newsweek.com/id/209164.
9. Hrdy (2001), p. 91.
10. "Scenes from a group marriage." By Laird Harrison. Salon.com. http://mobile.salon.com/mwt/feature/2008/06/04/open_marriage/index.html.
11. http://andrewsullivan.theatlantic.com/the_daily_dish/2009/01/ted-haggard-a-1.html.
12. McElvaine (2001), p. 339.
13. Perel (2006), p. 192.
14. Gould (2000), pp. 29–31.
15. After all, in the 1970s, *somebody* bought nearly four million copies of *Open Marriage*, by Nena and George O'Neill.
16. Perel (2006), pp. 192–194.
17. Burnham and Phelan (2000), p. 195.
18. Perel (2006), p. 197.
19. Bergstrand and Blevins Williams (2000).
20. Easton and Liszt (1997).
21. You can hear the press conference at www.thisamericanlife.org/Radio_Episode.aspx?episode=95.
22. We first learned of this striking sun/moon relation in Weil (1980), a fascinating book on the consciousness-altering potential of everything from solar eclipses to perfectly ripe mangoes.

REFERENCES AND SUGGESTED FURTHER READING

Abbott, E. (1999). *A History of Celibacy.* Cambridge, MA: Da Capo Press.

Abramson, P. R., and Pinkerton, S. D. (Eds.). (1995a). *Sexual Nature Sexual Culture.* Chicago: University of Chicago Press.

———(1995b). *With Pleasure: Thoughts on the Nature of Human Sexuality.* New York: Oxford University Press.

Acton, W. (1857/2008). *The Functions and Disorders of the Reproductive Organs in Childhood, Youth, Adult Age, and Advanced Live Considered in their Physiological, Social, and Moral Relations.* Charleston, SC: BiblioLife.

Adovasio, J. M., Soffer, O., and Page, J. (2007). *The Invisible Sex: Uncovering the True Roles of Women in Prehistory.* New York: Smithsonian Books.

Alexander, M. G., and Fisher, T. D. (2003). Truth and consequences: Using the bogus pipeline to examine sex differences in self-reported sexuality. *The Journal of Sex Research,* 40: 27–35.

Alexander, R. D. (1987). *The Biology of Moral Systems.* Chicago: Aldine.

Alexander, R. D., Hoogland, J. L., Howard, R. D., Noonan, K. M., and Sherman, P. W. (1979). Sexual dimorphisms and breeding systems in pinnepeds, ungulates, primates and humans. In N. Chagnon and W. Irons (Eds.), *Evolutionary Biology and Human Social Behavior: An Anthropological Perspective* (pp. 402–435). New York: Wadsworth.

Allen, M. L., and Lemmon, W. B. (1981). Orgasm in female primates. *American Journal of Primatology,* 1: 15–34.

Alvergne, A., and Lummaa, V. (2009). Does the contraceptive pill alter mate choice in humans? *Trends in Ecology and Evolution,* 24. In press—published online October 7, 2009.

Ambrose, S. (1998). Late Pleistocene human population bottlenecks, volcanic winter, and differentiation of modern humans. *Journal of Human Evolution* 34(6): 623–651.

Amos, W., and Hoffman, J. I. (2009). Evidence that two main bottleneck events shaped modern human genetic diversity. *Proceedings of the Royal*

Society B. Published online before print October 7, 2009, doi:10.1098/rspb.2009.1473.

Anderson, M., Hessel, J., and Dixson, A. F. (2004). Primate mating systems and the evolution of immune response. *Journal of Reproductive Immunology*, 61: 31–38.

Angier, N. (1995). *The Beauty of the Beastly: New Views of the Nature of Life*. New York: Houghton Mifflin.

———(1999). *Woman: An Intimate Geography*. New York: Virago.

Anokhin, A. P., Golosheykin, S., Sirevaag, E., Kristjansson, S., Rohrbaugh, J. W., and Heath, A. C. (2006). Rapid discrimination of visual scene content in the human brain. *Brain Research*, doi:10.1016/j.brainres.2006.03.108, available online May 18, 2006.

Ardrey, R. (1976). *The Hunting Hypothesis*. New York: Athenaeum.

Axelrod, R. (1984). *The Evolution of Cooperation*. New York: Basic Books.

Bagemihl, B. (1999). *Biological Exuberance: Animal Homosexuality and Natural Diversity*. New York: St. Martin's Press.

Baker, R. R. (1996). *Sperm Wars: The Science of Sex*. New York: Basic Books.

Baker, R. R., and Bellis, M. (1995). *Human Sperm Competition*. London: Chapman Hall.

Barash, D. P. (1977). *Sociobiology and Behavior*. Amsterdam: Elsevier.

Barash, D. P., and Lipton, J. E. (2001). *The Myth of Monogamy: Fidelity and Infidelity in Animals and People*. New York: W. H. Freeman.

Barkow, J. H. (1984). The distance between genes and culture. *Journal of Anthropological Research*, 40: 367–379.

Barkow, J. H., Cosmides, L., and Tooby, J. (Eds.). (1992). *The Adapted Mind: Evolutionary Psychology and the Generation of Culture*. New York: Oxford University Press.

Barlow, C. (Ed.). (1984). *Evolution Extended: Biological Debates on the Meaning of Life*. Cambridge, MA: MIT Press.

Barlow, N. (Ed.). (1958). *The Autobiography of Charles Darwin*. New York: Harcourt Brace.

Barratt, C. L. R., Kay, V., and Oxenham, S. K. (2009). The human spermatozoon—a stripped down but refined machine. *Journal of Biology*, 8: 63. http://jbiol.com/content/8/7/63.

Bateman, A. J. (1948). Intra-sexual selection in *Drosophila*. *Heredity*, 2: 349–368.

Batten, M. (1992). *Sexual Strategies: How Females Choose Their Mates*. New York: Putnam.

Baumeister, R. F. (2000). Gender differences in erotic plasticity: The female sex drive as socially flexible and responsive. *Psychological Bulletin,* 126: 347–374.

Beach, F. (Ed.). (1976). *Human Sexuality in Four Perspectives.* Baltimore: Johns Hopkins University Press.

Bean, L. J. (1978). Social organization. In R. Heizer (Ed.), *Handbook of North American Indians,* Vol. 8: *California* (pp. 673–682). Washington, D.C.: Smithsonian Institution Press.

Beckerman, S., and Valentine, P. (Eds.). (2002). *Cultures of Multiple Fathers: The Theory and Practice of Partible Paternity in Lowland South America.* Gainesville: University Press of Florida.

Bellis, M. A., and Baker, R. R. (1990). Do females promote sperm competition: Data for humans. *Animal Behaviour,* 40: 997–999.

Belliveau, J. (2006). *Romance on the Road: Travelling Women Who Love Foreign Men.* Baltimore, MD: Beau Monde Press.

Behar, D. M., et al. (2008). The dawn of human matrilineal diversity. *The American Journal of Human Genetics,* 82: 1130–1140.

Bergstrand, C., and Blevins Williams, J. (2000). Today's Alternative Marriage Styles: The Case of Swingers. *Electronic Journal of Human Sexuality*: Annual. Online: http://findarticles.com/p/articles/mi_6896/is_3/ai_n28819761/?tag=content;col1.

Berkowitz, B., and Yager-Berkowitz, S. (2008). *He's Just Not Up For It Anymore: Why Men Stop Having Sex and What You Can Do About It.* New York: William Morrow.

Berman, M. (2000). *Wandering God: A Study in Nomadic Spirituality.* Albany: State University of New York Press.

Bernard, J. (1972/1982). *The Future of Marriage.* New Haven: Yale University Press.

Betzig, L. (1982). Despotism and differential reproduction: A cross-cultural correlation of conflict asymmetry, hierarchy and degree of polygyny. *Ethology and Sociobiology,* 3: 209–221.

———(1986). *Despotism and Differential Reproduction: A Darwinian View of History.* New York: Aldine.

———(1989). Causes of conjugal dissolution: A cross-cultural study. *Current Anthropology,* 30: 654–676.

Birkhead, T. (2000). *Promiscuity: An Evolutionary History of Sperm Competition and Sexual Conflict.* New York: Faber and Faber.

———(2002). Postcopulatory sexual selection. *Nature Reviews: Genetics,* 3: 262–273. www.nature.com/reviews/genetics.

Blount, B. G. (1990). Issues in bonobo (*Pan paniscus*) sexual behavior. *American Anthropologist*, 92: 702–714.

Blum, D. (1997). *Sex on the Brain: The Biological Differences Between Men and Women*. New York: Viking.

Blurton Jones, N., Hawkes, K., and O'Connell, J. F. (2002). Antiquity of postreproductive life: Are there modern impacts on hunter-gatherer postreproductive life spans? *American Journal of Human Biology*, 14: 184–205.

Bodley, J. (2002). *Power of Scale: A Global History Approach (Sources and Studies in World History)*. Armonk, NY: M. E. Sharpe.

Boehm, C. H. (1999). *Hierarchy in the Forest: The Evolution of Egalitarian Behavior.* Cambridge, MA: Harvard University Press.

Bogucki , P. (1999). *The Origins of Human Society.* Malden, MA: Blackwell.

Borofsky, R. (2005). *Yanomami: The Fierce Controversy and What We Can Learn From It.* University of California Press.

Borries, C., Launhardt, K., Epplen, C., Epplen, J. T., and Winkler, P. (1999). Males as infant protectors in Hanuman langurs (*Presbytis entellus*) living in multimale groups—defense pattern, paternity and sexual behaviour. *Behavioral Ecology and Sociobiology*, 46: 350–356.

Bowlby, J. (1992). *Charles Darwin: A New Life.* New York: Norton.

Boyd, R., and Silk, J. (1997). *How Humans Evolved.* New York: Norton.

Boyle, T. C. (1993). *The Road to Wellville.* New York: Viking.

Boysen, S. T., and Himes, G. T. (1999). Current issues and emergent theories in animal cognition. *Annual Reviews in Psychology*, 50: 683–705.

Brizendine, L. (2006). *The Female Brain*. New York: Morgan Road Books.

Brown, D. (1970/2001). *Bury My Heart at Wounded Knee: An Indian History of the American West*. New York: Holt Paperbacks.

Bruhn, J. G., and Wolf, S. (1979). *The Roseto Story: An Anatomy of Health.* University of Oklahoma Press.

Buller, D. J. (2005). *Adapting Minds: Evolutionary Psychology and the Persistent Quest for Human Nature*. Cambridge, MA: The MIT Press.

Bullough, V. L. (1994). *Science in the Bedroom: A History of Sex Research.* New York: HarperCollins.

Burch, E. S., Jr., and Ellanna, L. J. (Eds.) (1994). *Key Issues in Hunter-Gatherer Research.* Oxford, England: Berg.

Burnham, T., and Phelan, J. (2000). *Mean Genes: From Sex to Money to Food: Taming Our Primal Instincts*. Cambridge, MA: Perseus.

Buss, D. M. (1989). Sex differences in human mate preferences: Evolutionary hypotheses testing in 37 cultures. *Behavioral and Brain Sciences*, 12: 1–49.

———(1994). *The Evolution of Desire: Strategies of Human Mating.* New York: Basic Books.

———(2000). *The Dangerous Passion: Why Jealousy Is as Necessary as Love and Sex.* New York: The Free Press.

———(2005). *The Murderer Next Door: Why the Mind Is Designed to Kill.* New York: Penguin Press.

Buss, D. M., Larsen, R. J., Westen, D., and Semmelroth, J. (1992). Sex differences in jealousy: Evolution, physiology and psychology. *Psychological Science,* 3: 251–255.

Buss, D. M., and Schmitt, D. P. (1993). Sexual strategies theory: An evolutionary perspective on human mating. *Psychological Review,* 100: 204–232.

Caesar, J. (2008). *The Gallic Wars: Julius Caesar's Account of the Roman Conquest of Gaul.* St. Petersburg, FL: Red and Black Publishers.

Cassini, M. H. (1998). Inter-specific infanticide in South American otariids. *Behavior,* 135: 1005–1012.

Caswell, J. L., et al. (2008). Analysis of chimpanzee history based on genome sequence alignments. *PLoS Genetics,* April; 4(4): e1000057. Online: http://www.plosgenetics.org/article/info%3Adoi%2F10.1371%2Fjournal.pgen.1000057.

Caton, H. (1990). *The Samoa Reader: Anthropologists Take Stock.* Lanham, MD: University Press of America.

Chagnon, N. (1968). *Yanomamö: The Fierce People.* New York: Holt, Rinehart and Winston.

Chapman, A. R., and Sussman, R. W. (Eds.). (2004.) *The Origins and Nature of Sociality.* Piscataway, NJ: Aldine Transaction.

Cherlin, A. J. (2009). *The Marriage-Go-Round: The State of Marriage and the Family in America Today.* New York: Knopf.

Chernela, J. M. (2002). Fathering in the northwest Amazon of Brazil. In S. Beckerman and P. Valentine (Eds.), *Cultures of Multiple Fathers: The Theory and Practice of Partible Paternity in Lowland South America* (pp. 160–177). Gainesville: University Press of Florida.

Chivers, M. L., Seto, M. C., and Blanchard, R. (2007). Gender and sexual orientation differences in sexual response to the sexual activities versus the gender of actors in sexual films. *Journal of Personality and Social Psychology,* 93: 1108–1121.

Clark, G. (1997). Aspects of early hominid sociality: An evolutionary perspective. In C. Barton and G. Clark (Eds.), *Rediscovering Darwin: Evo-*

lutionary Theory and Archaeological Explanation. Archaeological Papers of the American Anthropological Association, 7: 209–231.

Clark, R. D., and Hatfield, E. (1989). Gender differences in receptivity to sexual offers. *Journal of Psychology & Human Sexuality*, 2: 39–55.

Cochran, G., and Harpending, H. (2009). *The 10,000 Year Explosion: How Civilization Accelerated Human Evolution*. New York: Basic Books.

Cohen, S., et al. (2009). Sleep habits and susceptibility to the common cold. *Archives of Internal Medicine*, 169: 62.

Corning, P. (1994). The synergism hypothesis: A theory of progressive evolution. In C. Barlow (Ed.), *Evolution Extended: Biological Debates on the Meaning of Life* (pp. 110–118). Cambridge, MA: MIT Press.

Cosmides, L., and Tooby, J. (1987). From evolution to behavior: Evolutionary psychology as the missing link. In J. Dupree (Ed.), *The Latest on the Best: Essays on Evolution and Optimality* (pp. 227–306). Cambridge, MA: MIT Press.

Counts, D. E. A., and Counts, D. R. (1983). Father's water equals mother's milk: The conception of parentage in Kaliai, West New Guinea. *Mankind*, 14: 45–56.

Coventry, M. (October/November 2000). Making the cut: It's a girl! . . . or is it? When there's doubt, why are surgeons calling the shots? *Ms. Magazine*. Retrieved July 2, 2002 from http://www.msmagazine.com/oct00/makingthecut.html.

Crocker, W. H. (2002). Canela "other fathers": Partible paternity and its changing practices. In S. Beckerman and P. Valentine (Eds.), *Cultures of Multiple Fathers: The Theory and Practice of Partible Paternity in Lowland South America* (pp. 86–104). Gainesville: University Press of Florida.

Crocker, W. H., and Crocker, J. G. (2003). *The Canela: Kinship, Ritual and Sex in an Amazonian Tribe (Case Studies in Cultural Anthropology)*. Florence, KY: Wadsworth.

Dabbs, J. M., Jr., Carr, T. S., Frady, R. L., and Riad, J. K. (1995). Testosterone, crime and misbehavior among 692 male prison inmates. *Personality and Individual Differences*, 18: 627–633.

Dabbs, J. M., Jr., Jurkovic, G., and Frady, R. L. (1991). Salivary testosterone and cortisol among late adolescent male offenders. *Journal of Abnormal Child Psychology*, 19: 469–478.

Daniels, D. (1983). The evolution of concealed ovulation and self-deception. *Ethology and Sociobiology*, 4: 69–87.

Darwin, C. (1859). *On the Origin of Species by Means of Natural Selection*. London: John Murray.

———(1871/2007). *The Descent of Man and Selection in Relation to Sex*. New York: Plume.

Davenport, W. H. (1965). Sexual patterns and their regulation in a society of the southwest Pacific. In Beach (Ed.), *Sex and Behavior*, pp. 161–203.

Dawkins, R. (1976). *The Selfish Gene*. New York: Oxford University Press.

———(1998). *Unweaving the Rainbow: Science, Delusion and the Appetite for Wonder.* Boston: Houghton Mifflin.

DeMeo, J. (1998). *Saharasia: The 4000 B.C.E. Origins of Child Abuse, Sex-repression, Warfare and Social Violence, in the Deserts of the Old World*. Eugene, OR: Natural Energy Works.

Desmond, A., and Moore, J. (1994). *Darwin: The Life of a Tormented Evolutionist.* New York: Warner Books.

DeSteno, D., and Salovey, P. (1996). Evolutionary origins of sex differences in jealousy? Questioning the "fitness" of the model. *Psychological Science*, 7: 367–372.

de Waal, F. (1995). Bonobo sex and society: The behavior of a close relative challenges assumptions about male supremacy in human evolution. *Scientific American* (March): 82–88.

———(1996). *Good Natured: The Origins of Right and Wrong in Humans and Other Animals*. Cambridge, MA: Harvard University Press.

———(1998). *Chimpanzee Politics: Power and Sex among the Apes*. Baltimore, MD: Johns Hopkins University Press. (Original work published 1982.)

———(2001a). *The Ape and the Sushi Master: Cultural Reflections of a Primatologist*. New York: Basic Books.

———(2001b). Apes from Venus: Bonobos and human social evolution. In F. de Waal (Ed.), *Tree of Origin: What Primate Behavior Can Tell Us About Human Social Evolution* (pp. 39–68). Cambridge, MA: Harvard University Press.

———(2001c). (Ed.). *Tree of Origin: What Primate Behavior Can Tell Us About Human Social Evolution*. Cambridge, MA: Harvard University Press.

———(2005a). *Our Inner Ape: The Best and Worst of Human Nature*. London: Granta Books.

———(2005b). Bonobo sex and society. *Scientific American* online issue, February, pp. 32–38.

———(2009). *The Age of Empathy: Nature's Lessons for a Kinder Society*. New York: Harmony Books.

de Waal, F., and Johanowicz, D. L. (1993). Modification of reconciliation behavior through social experience: An experiment with two macaque species. *Child Development* 64: 897–908.

de Waal, F., and Lanting, F. (1998). *Bonobo: The Forgotten Ape*. Berkeley: University of California Press.

Dewsbury, D. A. (1981). Effects of novelty on copulatory behavior: The Coolidge effect and related phenomena. *Psychological Bulletin,* 89: 464–482.

Diamond, J. (1986). Variation in human testis size. *Nature*, 320: 488.

———(1987). The worst mistake in the history of the human race. *Discover,* May.

———(1991). *The Rise and Fall of the Third Chimpanzee: How Our Animal Heritage Affects the Way We Live.* London: Vintage.

———(1997). *Guns, Germs and Steel: The Fates of Human Societies.* New York: Norton.

———(2005). *Collapse: How Societies Choose to Fail or Succeed.* New York: Viking.

Diamond, L. M. (2008). *Sexual Fluidity: Understanding Women's Love and Desire.* Cambridge, MA: Harvard University Press.

Dillehay, T. D., et al. (2008). Monte Verde: Seaweed, food, medicine and the peopling of South America. *Science,* 320 (5877): 784–786.

Dindyal, S. (2004). The sperm count has been decreasing steadily for many years in Western industrialised countries: Is there an endocrine basis for this decrease? *The Internet Journal of Urology,* 2(1).

Dixson, A. F. (1998). *Primate Sexuality: Comparative Studies of the Prosimians, Monkeys, Apes and Human Beings.* New York: Oxford University Press.

Dixson, A. F., and Anderson, M. (2001). Sexual selection and the comparative anatomy of reproduction in monkeys, apes and human beings. *Annual Review of Sex Research,* 12: 121–144.

———(2002). Sexual selection, seminal coagulation and copulatory plug formation in primates. *Folia Primatologica,* 73: 63–69.

Drucker, D. (2004). *Invent Radium or I'll Pull Your Hair: A Memoir.* Chicago: University of Chicago Press.

Druckerman, P. (2008). *Lust in Translation: Infidelity from Tokyo to Tennessee.* New York: Penguin Two.

Dunbar, R. I. M. (1992). Neocortex size as a constraint on group size in primates. *Journal of Human Evolution,* 22: 469–493.

Dunbar, R. I. M. (1993). Coevolution of neocortical size, group size and language in humans. *Behavioral and Brain Sciences,* 16 (4): 681–735.

Easton, D., and Liszt, C. A. (1997). *The Ethical Slut: A Guide to Infinite Sexual Possibilities.* San Francisco, CA: Greenery Press.

Eaton, S., and Konner, M. (1985). Paleolithic nutrition: A consideration of its nature and current implications. *New England Journal of Medicine,* 312: 283–289.

Eaton, S., Konner, M., and Shostak, M. (1988). Stone agers in the fast lane: Chronic degenerative disease in evolutionary perspective. *American Journal of Medicine,* 84: 739–749.

Eaton, S., Shostak, M., and Konner, M. (1988). *The Paleolithic Prescription: A Program of Diet & Exercise and a Design for Living.* New York: Harper & Row.

Eberhard, W. G. (1985). *Sexual Selection and Animal Genitalia.* Cambridge, MA: Harvard University Press.

Eberhard, W. G. (1996). *Female Control: Sexual Selection by Cryptic Female Choice.* Princeton, NJ: Princeton University Press.

Edgerton, R. B. (1992). *Sick Societies: Challenging the Myth of Primitive Harmony.* New York: The Free Press.

Ehrenberg, M. (1989). *Women in Prehistory.* London: British Museum Publications.

Ehrlich, P. R. (2000). *Human Natures: Genes, Cultures, and the Human Prospect.* New York: Penguin.

Ellison, P. T., et al. (2009). *Endocrinology of Social Relationships.* Cambridge, MA: Harvard University Press.

Elwin, V. (1968). *Kingdom of the Young.* Bombay: Oxford University Press.

Erikson, P. (1993). A onomástica matis é amazônica [Naming rituals among the Matis of the Amazon]. In E. Viveiros de Castro and M. Carneiro da Cuhna (Eds.), *Amazônia: Etnologia e história indígena* (pp. 323–338). São Paulo: Nucléo de História Indíena et do Indigenismo, USP/FAPESP.

——— (2002). Several fathers in one's cap: Polyandrous conception among the Panoan Matis (Amazonas, Brazil). In S. Beckerman and P. Valentine (Eds.), *Cultures of Multiple Fathers: The Theory and Practice of Partible Paternity in Lowland South America* (pp. 123–136). Gainesville: University Press of Florida.

Ewen, S. (1976/2001). *Captains of Consciousness: Advertising and the Social Roots of the Consumer Culture.* New York: Basic Books.

Fagan, B. (2004). *The Long Summer: How Climate Changed Civilization.* New York: Basic Books.

Fedigan, L. M., and Strum, S. C. (1997). Changing images of primate societies. *Current Anthropology,* 38: 677–681.

Feinstein, D., and Krippner, S. (2007). *The Mythic Path: Discovering the Guiding Stories of Your Past—Creating a Vision for Your Future.* Fulton, CA: Elite Books.

Ferguson, B. (1995). *Yanomami Warfare: A Political History.* Santa Fe, NM: School of American Research Press.

———(2000). *War in the Tribal Zone: Expanding States and Indigenous Warfare.* Santa Fe, NM: SAR Press.

———(2003). The birth of war. *Natural History*, July/August: 28–34.

Ferraro, G., Trevathan, W., and Levy, J. (1994). *Anthropology: An Applied Perspective.* Minneapolis/St. Paul, MN: West Publishing Company.

Fish, R. C. (2000). *The Clitoral Truth: The Secret World at Your Fingertips.* New York: Seven Stories Press.

Fisher, H. E. (1982). *The Sex Contract: The Evolution of Human Behavior.* New York: William Morrow.

——— (1989). Evolution of human serial pairbonding. *American Journal of Physical Anthropology*, 78: 331–354.

———(1992). *Anatomy of Love.* New York: Fawcett Columbine.

——— (2004). *Why We Love: The Nature and Chemistry of Romantic Love.* New York: Henry Holt.

Fisher, M., et al. (2009). Imact of relational proximity on distress from infidelity. *Evolutionary Psychology*, 7(4): 560–580.

Flanagan, C. (2009). Is there hope for the American Marriage? *Time*, July 2. http://www.time.com/time/nation/article/0,8599,1908243-1,00.html.

Fleming, J. B. (1960). Clitoridectomy: The disastrous downfall of Isaac Baker Brown, F.R.C.S. (1867). *Journal of Obstetrics and Gynaecology of the British Empire*, 67: 1017–1034.

Foley, R. (1996). The adaptive legacy of human evolution: A search for the environment of evolutionary adaptiveness. *Evolutionary Anthropology*, 4: 194–203.

Ford, C. S., and Beach, F. (1952). *Patterns of Sexual Behavior.* Westport, CT: Greenwood Press.

Fordney-Settlage, D. (1981). A review of cervical mucus and sperm interactions in humans. *International Journal of Fertility*, 26: 161–169.

Fortenberry, D. J. (2005). The limits of abstinence-only in preventing sexually transmitted infections. *Journal of Adolescent Health*, 36: 269–357.

Fouts, R., with Mills, S. T. (1997). *Next of Kin: My Conversations with Chimpanzees.* New York: Avon Books.

Fox, C. A., Colson, R. H., and Watson, B. W. (1982). Continuous measurement of vaginal and intra-uterine pH by radio-telemetry during human coitus. In Z. Hoch and H. L. Lief (Eds.), *Sexology* (pp. 110–113). Amsterdam: Excerpta Medica.

Fox, R. (1997). *Conjectures & Confrontations: Science, Evolution, Social Concern.* Somerset, NJ: Transaction.

Fowles, J. (1969). *The French Lieutenant's Woman.* New York: Signet.

Freeman, D. (1983). *Margaret Mead and Samoa: The Making and Unmaking of an Anthropological Myth.* Cambridge, MA: Harvard University Press.

Friedman, D. M. (2001). *A Mind of Its Own: A Cultural History of the Penis.* New York: The Free Press.

Fromm, E. (1973). *The Anatomy of Human Destructiveness.* New York: Holt, Rinehart and Winston.

Fry, D. (2009). *Beyond War: The Human Potential for Peace.* New York: Oxford University Press.

Gagneaux, P., and Boesch, C. (1999). Female reproductive strategies, paternity and community structure in wild West African chimpanzees. *Animal Behaviour,* 57: 19–32.

Gallup, G. G., Jr. (2009). On the origin of descended scrotal testicles: The activation hypothesis. *Evolutionary Psychology,* 7: 517–526. Available online at http://www.epjournal.net.

Gallup, G. G., Jr., and Burch, R. L. (2004). Semen displacement as a sperm competition strategy in humans. *Evolutionary Psychology,* 2: 12–23. Available online at http://www.epjournal.net.

Gallup, G. G., Jr., Burch, R. L., and Platek, S. M. (2002). Does semen have antidepressant properties? *Archives of Sexual Behavior,* 31: 289–293.

Gangestad, S. W., Bennett, K., and Thornhill, R. (2001). A latent variable model of developmental instability in relation to men's sexual behavior. *Proceedings of the Royal Society of London,* 268: 1677–1684.

Gangestad, S. W., and Thornhill, R. (1998). Menstrual cycle variation in women's preferences for the scent of symmetrical men. *Proceedings of the Royal Society of London,* 265: 927–933.

Gangestad, S. W., Thornhill, R., and Yeo, R. A. (1994). Facial attractiveness, developmental stability and fluctuating symmetry. *Ethology and Sociobiology,* 15: 73–85.

Ghiglieri, M. P. (1999). *The Dark Side of Man: Tracing the Origins of Male Violence.* Reading, MA: Helix Books.

Gibson, P. (1989). Gay and lesbian youth suicide, in Fenleib, Marcia R. (Ed.), *Report of the Secretary's Task Force on Youth Suicide,* United States Government Printing Office, ISBN 0160025087.

Gladwell, M. (2002). *The Tipping Point: How Little Things Can Make a Big Difference*. New York: Back Bay Books.

——— (2008). *Outliers: The Story of Success*. New York: Little, Brown and Company.

Glass, D. P., and Wright, T. L. (1985). Sex differences in type of extramarital involvement and marital dissatisfaction. *Sex Roles*, 12: 1101–1120.

Goldberg, S. (1993). *Why Men Rule: A Theory of Male Dominance*. Chicago: Open Court.

Good, K., with Chanoff, D. (1991). *Into the Heart: One Man's Pursuit of Love and Knowledge Among the Yanomama*. Leicester, England: Charnwood.

Goodall, J. (1971). *In the Shadow of Man*. Glasgow: Collins.

———(1991). *Through a Window: Thirty Years with the Chimpanzees of Gombe*. London: Penguin.

Goodman, M., et al. (1998). Toward a phylogenic classification of primates based on DNA evidence complemented by fossil evidence. *Molecular Phylogenics and Evolution*, 9: 585–598.

Gould, S. J. (1980). *Ever since Darwin: Reflections in Natural History*. New York: Norton.

———(1981). *The Mismeasure of Man*. New York: Norton.

———(1991). Exaptation: A crucial tool for an evolutionary psychology. *Journal of Social Issues*, 47(3): 43–65.

———(1997). Darwinian fundamentalism. *New York Review of Books*, pp. 34–37. Retrieved December 12, 2002 from http://www.nybooks.com/articles/1151.

Gould, S. J., and Lewontin, R. C. (1979). The spandrels of San Marco and the Panglossian paradigm: A critique of the adaptionist programme. *Proceedings of the Royal Society of London*, 205: 581–598.

Gould, S. J., and Vrba, E. S. (1982). Exaptation—a missing term in the science of form. *Paleobiology*, 8: 4–15.

Gould, T. (2000). *The Lifestyle: A Look at the Erotic Rites of Swingers*. Buffalo, NY: Firefly Books.

Gowdy, J. (Ed.). (1998). *Limited Wants, Unlimited Means: A Reader on Hunter-gatherer Economics and the Environment*. Washington, DC: Island Press.

Gray, P. B., Kahlenberg, S. M., Barrett, E. S., Lipson, S. F., and Ellison, P. T. (2002). Marriage and fatherhood are associated with lower testosterone in males. *Evolution and Human Behavior*, 23(3): 193–201.

Gray, P. B., Parkin, J. C., and Samms-Vaughan, M. E. (1997). Hormonal correlates of human paternal interactions: A hospital-based investigation in urban Jamaica. *Hormones and Behavior*, 52: 499–507.

Gregor, T. (1985). *Anxious Pleasures: The Sexual Lives of an Amazonian People.* Chicago: University of Chicago Press.

Hamilton, W. D. (1964). The genetic evolution of social behavior. Parts I and II. *Journal of Theoretical Biology,* 7: 1–52.

———(2001). *The Narrow Roads of Gene Land.* New York: Oxford University Press.

Hamilton, W. J., and Arrowood, P. C. (1978). Copulatory vocalizations of Chacma baboons (*Papio ursinus*), gibbons (*Hylobates hoolock*) and humans. *Science,* 200: 1405–1409.

Harcourt, A. H. (1997). Sperm competition in primates. *American Naturalist,* 149: 189–194.

Harcourt, A. H., and Harvey, P. H. (1984). Sperm competition, testes size and breeding systems in primates. In R. Smith (Ed.), *Sperm Competition and the Evolution of Animal Mating Systems* (pp. 589–659). New York: Academic Press.

Hardin, G. (1968). The tragedy of the commons. *Science,* 131: 1292–1297.

Harper, M. J. K. (1988). Gamete and zygote transport. In E. Knobil and J. Neill (Eds.), *The Physiology of Reproduction* (pp. 103–134). New York: Raven Press.

Harris, C. (2000). Psychophysiological responses to imagined infidelity: The specific innate modular view of jealousy reconsidered. *Journal of Personality and Social Psychology,* 78: 1082–1091.

Harris, C., and Christenfeld, N. (1996). Gender, jealousy and reason. *Psychological Science,* 7: 364–366.

Harris, M. (1977). *Cannibals and Kings: The Origins of Cultures.* New York: Random House.

———(1980). *Cultural Materialism: The Struggle for a Science of Culture.* New York: Vintage Books.

———(1989). *Our Kind: Who We Are, Where We Came From, Where We Are Going.* New York: Harper & Row.

———(1993). The evolution of human gender hierarchies: A trial formulation. In B. Miller (Ed.), *Sex and Gender Hierarchies* (pp. 57–79). Cambridge, England: Cambridge University Press.

Hart, D., and Sussman, R. W. (2005). *Man the Hunted: Primates, Predators, and Human Evolution.* New York: Westview Press.

Harvey, P. H., and May, R. M. (1989). Out for the sperm count. *Nature,* 337: 508–509.

Haselton, M. G., et al. (2007). Ovulatory shifts in human female ornamentation:

Near ovulation, women dress to impress. *Hormones and Behavior,* 51: 40–45. www.sscnet.ucla.edu/comm/haselton/webdocs/dress_to_impress.pdf.

Hassan, F. A. (1980). The growth and regulation of human population in prehistoric times. In Cohen, M. N., Malpass, R. S., and Klein, H. G. (Eds.), *Biosocial Mechanisms of Population Regulation* (pp. 305–319). New Haven, CT: Yale University Press.

Hawkes, K. (1993). Why hunter-gatherers work. *Current Anthropology,* 34: 341–361.

Hawkes, K., O'Connell, J. F., and Blurton Jones, N. G. (2001a). Hadza meat sharing. *Evolution and Human Behavior,* 22: 113–142.

———(2001b). Hadza hunting and the evolution of nuclear families. *Current Anthropology,* 42: 681–709.

Heinen, H. D., and Wilbert, W. (2002). Parental uncertainty and ritual kinship among the Warao. In S. Beckerman and P. Valentine (Eds.), *Cultures of Multiple Fathers: The Theory and Practice of Partible Paternity in Lowland South America* (pp. 210–220). Gainesville: University Press of Florida.

Henderson, V. W., and Hogervorst, E. (2004). Testosterone and Alzheimer disease: Is it men's turn now? *Neurology,* 62: 170–171.

Henrich, J., et al. (2005). "Economic man" in cross-cultural perspective: Behavioral experiments in 15 small-scale societies. *Behavioral and Brain Sciences,* 28: 795–855.

Highwater, J. (1990). *Myth and Sexuality.* New York: New American Library.

Hill, K., and Hurtado, M. (1996). *Aché Life History: The Ecology and Demography of a Foraging People.* New York: Aldine de Gruyter.

Hite, S. (1987). *Women and Love: A Cultural Revolution in Progress.* New York: Knopf.

———(1989). *The Hite Report: A Nationwide Study of Female Sexuality.* New York: Dell.

Hobbes, T. (1991). *Leviathan.* Cambridge, England: Cambridge University Press. (Original work published 1651.)

Holmberg, A. R. (1969). *Nomads of the Long Bow: The Siriono of Eastern Bolivia.* New York: The Natural History Press.

Horne, B. D., et al. (2008). Usefulness of routine periodic fasting to lower risk of coronary artery disease in patients undergoing coronary angiography. *American Journal of Cardiology,* 102(7): 814–819.

Houghton, W. E. (1957). *The Victorian Frame of Mind, 1830–1870.* New Haven, CT: Yale University Press.

Hrdy, S. B. (1979). Infanticide among animals: A review, classification and

examination of the implications for the reproductive strategies of females. *Ethology and Sociobiology,* 1: 13–40.
———(1988). The primate origins of human sexuality. In R. Bellig and G. Stevens (Eds.), *The Evolution of Sex* (pp. 101–136). San Francisco: Harper and Row.
———(1996). Raising Darwin's consciousness: Female sexuality and the prehominid origins of patriarchy. *Human Nature,* 8(1):1–49.
———(1999a). *The Woman That Never Evolved.* Cambridge, MA: Harvard University Press. (Original work published 1981.)
———(1999b). *Mother Nature: A History of Mothers, Infants and Natural Selection.* Boston: Pantheon Books.
———(2009). *Mothers and Others: The Evolutionary Origins of Mutual Understanding.* Cambridge, MA: Harvard University Press.
Hua, C. (2001). *A Society Without Fathers or Husbands: The Na of* China. New York: Zone Books.
Human Genome Project. (2002). Retrieved November 11, 2002 from http://www.ornl.gov/hgmis.
Ingold, T., Riches, D., and Woodburn, J. (Eds.) (1988a). *Hunters and Gatherers: History, Evolution and Social Change* (Vol. 1). Oxford, England: Berg.
———(1988b). *Hunters and Gatherers: Property, Power and Ideology* (Vol. 2). Oxford, England: Berg.
Isaac, G. (1978). The food sharing behavior of protohuman hominids. *Scientific American,* 238(4): 90–108.
Janus, S. S., and Janus, C. L. (1993). *The Janus Report on Sexual Behavior.* New York: Wiley.
Jaynes, J. (1990). *The Origins of Consciousness in the Breakdown of the Bicameral Mind.* Boston: Houghton Mifflin. (Original work published 1976.)
Jethá, C., and Falcato, J. (1991). A mulher e as DTS no distrito de Marracuene [Women and Sexually Transmitted Diseases in the Marracuene district]. *Acção SIDA 9*, Brochure.
Jiang, X., Wang, Y., and Wang, Q. (1999). Coexistence of monogamy and polygyny in black-crested gibbon. *Primates,* 40(4): 607–611.
Johnson, A. W., and Earle, T. (1987). *The Evolution of Human Societies: From Foraging Group to Agrarian State.* Palo Alto, CA: Stanford University Press.
Jones, S., Martin, R. D., and Pilbeam, D. (Eds.) (1992). *The Cambridge Encyclopedia of Human Evolution.* Cambridge, UK: Cambridge University Press.

Jung, C. G. (1976). *The Symbolic Life: The Collected Works* (Vol. 18, Bolligen Series). Princeton, NJ: Princeton University Press.

Kanazawa, S. (2007). The evolutionary psychological imagination: Why you can't get a date on a Saturday night and why most suicide bombers are Muslim. *Journal of Social, Evolutionary and Cultural Psychology,* 1(2): 7–17.

Kane, J. (1996). *Savages.* New York: Vintage.

Kano, T. (1980). Social behavior of wild pygmy chimpanzees (*Pan paniscus*) of Wamba: A preliminary report. *Journal of Human Evolution,* 9: 243–260.

——— (1992). *The Last Ape: Pygmy Chimpanzee Behavior and Ecology.* Palo Alto, CA: Stanford University Press.

Kaplan, H., Hill, K., Lancaster, J., and Hurtado, A. M. (2000). A theory of human life history evolution: Diet, intelligence and longevity. *Evolutionary Anthropology,* 9: 156–185.

Keeley, L. H. (1996). *War Before Civilization: The Myth of the Peaceful Savage.* New York: Oxford University Press.

Kelly, R. L. (1995). *The Foraging Spectrum: Diversity in Hunter-Gatherer Lifeways.* Washington, DC: Smithsonian Institution Press.

Kendrick, K. M., Hinton, M. R., Atkins, K., Haupt, M. A., and Skinner, J. D. (September 17, 1998). Mothers determine sexual preferences. *Nature,* 395: 229–230.

Kent, S. (1995). Unstable households in a stable Kalahari community in Botswana. *American Anthropologist,* 97: 39–54.

Kilgallon, S. J., and Simmons, L. W. (2005). Image content influences men's semen quality. *Biology Letters,* 1: 253–255.

Kingan, S. B., Tatar, M., and Rand, D. M. (2003). Reduced polymorphism in the chimpanzee semen coagulating protein, Semenogelin I. *Journal of Molecular Evolution,* 57: 159–169.

Kinsey, A. C., Pomeroy, W. B., and Martin, C. E. (1948). *Sexual Behavior in the Human Male.* Philadelphia: Saunders.

———(1953). *Sexual Behavior in the Human Female.* Philadelphia: Saunders.

Knight, C. (1995). *Blood Relations: Menstruation and the Origins of Culture.* New Haven, CT: Yale University Press.

Komisaruk, B. R., Beyer-Flores, C., and Whipple, B. (2006). *The Science of Orgasm.* Baltimore: The Johns Hopkins University Press.

Konner, M. (1982). *The Tangled Wing.* New York: Holt, Rinehart and Winston.

Knauft, B. (1987). Reconsidering violence in simple human societies: Homicide among the Gebusi of New Guinea. *Current Anthropology,* 28(4): 457–500.

———(2009). *The Gebusi: Lives Transformed in a Rainforest World*. New York: McGraw-Hill.

Krech, S. (1999). *The Ecological Indian: Myth and History*. New York: Norton.

Krieger, M. J. B., and Ross, K.G. (2002). Identification of a major gene regulating complex social behavior. *Science*, 295: 328–332.

Kuper, A. (1988). *The Invention of Primitive Society: Transformations of an Illusion*. London: Routledge.

Kundera, M. (1984). *The Unbearable Lightness of Being*. London: Faber and Faber.

Kuukasjärvi, S., Eriksson, C. J. P., Koskela, E., Mappes, T., Nissinen, K., and Rantala, M. J. (2004). Attractiveness of women's body odors over the menstrual cycle: The role of oral contraceptives and receiver sex. *Behavioral Ecology*, 15(4): 579–584.

Laan, E., Sonderman, J., and Janssen, E. (1995). Straight and lesbian women's sexual responses to straight and lesbian erotica: No sexual orientation effects. Poster session, 21st meeting of the International Academy of Sex Research, Provincetown, MA, September.

Laeng, B., and Falkenberg, L. (2007). Women's pupillary responses to sexually significant others during the hormonal cycle. *Hormones and Behavior*, 52: 520–530.

Ladygina-Kohts, N. N. (2002). *Infant Chimpanzee and Human Child: A Classic 1935 Comparative Study of Ape Emotions and Intelligence*. New York: Oxford University Press.

Lancaster, J. B., and Lancaster, C. S. (1983). Parental investment: The hominid adaptation. In D. J. Ortner (Ed.), *How Humans Adapt: A Biocultural Odyssey* (pp. 33–65). Washington, DC: Smithsonian Institution Press.

Larrick, J. W., Yost, J. A., Kaplan, J., King, G., and Mayhall, J. (1979). Patterns of health and disease among the Waorani Indians of eastern Ecuador. *Medical Anthropology*, 3(2): 147–189.

Laumann, E. O., Paik, A., and Rosen, R. C. (1999). Sexual dysfunction in the United States: Prevalence and predictors. *Journal of the American Medical Association*, 281: 537–544.

Lawler, R. R. (2009). Monomorphism, male-male competition, and mechanisms of sexual dimorphism. *Journal of Human Evolution*, 57: 321–325.

Lea, V. (2002). Multiple paternity among the M˜ebengokre (Kayopó, Jê) of central Brazil. In S. Beckerman and P. Valentine (Eds.), *Cultures of Multiple Fathers: The Theory and Practice of Partible Paternity in Lowland South America* (pp. 105–122). Gainesville: University Press of Florida.

Leacock, E. (1981). *Myths of Male Dominance: Collected Articles on Women Cross-Culturally.* New York: Monthly Review Press.

———(1998). Women's status in egalitarian society: Implications for social evolution. In J. Gowdy (Ed.), *Limited Wants, Unlimited Means: A Reader on Hunter-gatherer Economics and the Environment* (pp. 139–164). Washington, DC: Island Press.

LeBlanc, S. A., with Resgister, K.E. (2003). *Constant Battles: The Myth of the Peaceful, Noble Savage.* New York: St. Martin's Press.

Lee, R. B. (1968). What hunters do for a living, or, how to make out on scarce resources. In R. Lee and I. Devore (Eds.), *Man the Hunter* (pp. 30–48). Chicago: Aldine.

———(1969). !Kung bushman subsistence: An input-output analysis. In A. Vayde (Ed.), *Environment and Cultural Behavior* (pp. 73–94). Garden City, NY: Natural History Press.

———(1979). *The !Kung San: Men, Women and Work in a Foraging Society.* Cambridge, England: Cambridge University Press.

———(1998). Forward to J. Gowdy (Ed.), *Limited Wants, Unlimited Means: A Reader on Hunter-gatherer Economics and the Environment* (pp. ix–xii). Washington, DC: Island Press.

Lee, R. B., and Daly, R. (Eds.). (1999). *The Cambridge Encyclopedia of Hunters and Gatherers.* Cambridge, UK: Cambridge University Press.

Lee, R. B., and DeVore, I. (Eds.). (1968). *Man the Hunter.* Chicago: Aldine.

Le Jeune, P. (1897/2009). *Les relations des Jesuites. 1656–1657.* Toronto Public Library.

LeVay, S. (1994). *The Sexual Brain.* Cambridge, MA: The MIT Press.

Levine, L. W. (1996). *The Opening of the American Mind: Canons, Culture, and History.* Boston, MA: Beacon Press.

Levitin, D. J. (2009). *The World in Six Songs: How the Musical Brain Created Human Nature.* New York: Plume.

Lilla, M. (2007). *The Stillborn God: Religion, Politics and the Modern West.* New York: Knopf.

Lindholmer, C. (1973). Survival of human sperm in different fractions of split ejaculates. *Fertility and Sterility* 24: 521–526.

Lippa, R. A. (2007). The relation between sex drive and sexual attraction to men and women: A cross-national study of heterosexual, bisexual and homosexual men and women. *Archives of Sexual Behavior,* 36: 209–222.

Lishner, D. A., et al. (2008). Are sexual and emotional infidelity equally upsetting to men and women? Making sense of forced-choice responses. *Evolutionary Psychology*, 6(4): 667–675. Available online at http://www.epjournal.net.

Littlewood, I. (2003). *Sultry Climates: Travel and Sex.* Cambridge, MA: Da Capo Press.

Lovejoy, C. O. (1981). The origin of man. *Science*, 211: 341–350.

———(2009). Reexamining human origins in light of *Ardipithecus ramidus.* Science, 326: 74, 74e1–74e8.

Low, B. S. (1979). Sexual selection and human ornamentation. In N. A. Chagnon and W. Irons (Eds.), *Evolutionary Biology and Human Social Behavior* (pp. 462–487). Boston: Duxbury Press.

MacArthur, R. H., and Wilson, E. O. (1967). *Theory of Island Biogeography (Monographs in Population Biology*, Vol. 1). Princeton, NJ: Princeton University Press.

MacDonald, K. (1990). Mechanisms of sexual egalitarianism in Western Europe. *Ethology and Sociobiology*, 11: 195–238.

Macrides, F., Bartke, A., and Dalterio, S. (1975). Strange females increase plasma testosterone levels in male mice. *Science*, 189(4208): 1104–1106.

Maines, R. P. (1999). *The Technology of Orgasm: "Hysteria," the Vibrator and Women's Sexual Satisfaction*. Baltimore: Johns Hopkins University Press.

Malinowski, B. (1929). *The Sexual Life of Savages in North-Western Melanesia: An Ethnographic Account of Courtship, Marriage and Family Life Among the Natives of the Trobriand Islands, British New Guinea*. New York: Harcourt Brace.

———(1962). *Sex, Culture and Myth*. New York: Harcourt Brace.

Malkin, C. J., Pugh, P. J., Jones, R. D., Jones, T. H., and Channer, K. S. (2003). Testosterone as a protective factor against atherosclerosis—immunomodulation and influence upon plaque development and stability. *Journal of Endocrinology*, 178: 373–380.

Malthus, T. R. (1798). *An Essay on the Principle of Population: Or a View of Its Past and Present Effects on Human Happiness; with an Inquiry Into Our Prospects Respecting the Future Removal or Mitigation of the Evils which It Occasions*. London: John Murray. Full text: http://www.econlib.org/library/Malthus/malPlong.html.

Manderson, L., Bennett, L. R., and Sheldrake, M. (1999). Sex, social institutions and social structure: Anthropological contributions to the study of sexuality. *Annual Review of Sex Research*, 10: 184–231.

Margolis, J. (2004). *O: The Intimate History of the Orgasm*. New York: Grove Press.

Margulis, L., and Sagan, D. (1991). *Mystery Dance: On the Evolution of Human Sexuality.* New York: Summit Books.

Marshall, L. (1976/1998). Sharing, taking and giving: Relief of social tensions among the !Kung. In J. Gowdy (Ed.), *Limited Wants, Unlimited Means: A Reader on Hunter-gatherer Economics and the Environment* (pp. 65–85). Washington, DC: Island Press.

Martin, R. D., Winner, L. A., and Dettling, A. (1994). The evolution of sexual size dimorphism in primates. In R. V. Short and E. Balaban (Eds.), *The Differences Between the Sexes* (pp. 159–200). Cambridge, England: Cambridge University Press.

Masters, W., and Johnson, V. (1966). *Human Sexual Response.* Boston: Little, Brown.

Masters, W., Johnson, V., and Kolodny, R. (1995). *Human Sexuality.* Boston: Addison-Wesley.

McArthur, M. (1960). Food consumption and dietary levels of groups of aborigines living on naturally occurring foods. In C. P. Mountford (Ed.), *Records of the Australian-American Scientific Expedition to Arnhem Land,* Vol. 2: *Anthropology and Nutrition.* Melbourne, Australia: Melbourne University Press.

McCarthy, F. D., and McArthur, M. (1960). The food quest and the time factor in aboriginal economic life. In C. P. Mountford (Ed.), *Records of the Australian-American Scientific Expedition to Arnhem Land,* Vol. 2: *Anthropology and Nutrition.* Melbourne, Australia: Melbourne University Press.

McDonald, R. (1998). *Mr. Darwin's Shooter.* New York: Atlantic Monthly Press.

McElvaine, R. S. (2001). *Eve's Seed: Biology, the Sexes and the Course of History.* New York: McGraw-Hill.

McGrew, W. C., and Feistner, T. C. (1992). Two nonhuman primate models for the evolution of human food sharing: Chimpanzees and callitrichids. In J. Barkow, L. Cosmides, and J. Tooby (Eds.), *The Adapted Mind: Evolutionary Psychology and the Generation of Culture* (pp. 229–243). New York: Oxford University Press.

McNeil, L., Osborne, J., and Pavia, P. (2006). *The Other Hollywood: The Uncensored Oral History of the Porn Film Industry.* New York: It Books.

Mead, M. (1961). *Coming of Age in Samoa: A Psychological Study of Primitive Youth for Western Civilization.* New York: Morrow. (Original work published 1928.)

Menzel, P., and D'Aluisio, F. (1998). *Man Eating Bugs: The Art and Science of Eating Insects.* Berkeley, CA: Ten Speed Press.

Mill, J. S. (1874). On the Definition of Political Economy, and on the Method of Investigation Proper to It. *London and Westminster Review*, October 1836. In, *Essays on Some Unsettled Questions of Political Economy*, 2nd ed. London: Longmans, Green, Reader & Dyer.

Miller, G. (1998). How mate choice shaped human nature: A review of sexual selection and human evolution. In C. Crawford and D. Krebs (Eds.), *Handbook of evolutionary psychology: Ideas, issues, and applications* (pp. 87–129). Mahwah, NJ: Lawrence Erlbaum.

———(2000). *The Mating Mind: How Sexual Choice Shaped the Evolution of Human Nature*. New York: Doubleday.

Mitani, J., and Watts, D. (2001). Why do chimpanzees hunt and share meat? *Animal Behaviour*, 61: 915–924.

Mitani, J. C., Watts, D. P., and Muller, M. (2002). Recent developments in the study of wild chimpanzee behavior. *Evolutionary Anthropology*, 11: 9–25.

Mithen, S. (1996). *The Prehistory of the Mind*. London: Thames and Hudson.

———(2004). *After the Ice: A Global Human History*. Cambridge, MA: Harvard University Press.

———(2007). Did farming arise from a misapplication of social intelligence? *Philosophical Transactions of the Royal Society B*, 362: 705–718.

Moore, H. D. M., Martin, M., and Birkhead, T. R. (1999). No evidence for killer sperm or other selective interactions between human spermatozoa in ejaculates of different males in vitro. *Procedings of the Royal Society of London B*, 266: 2343–2350.

Monaghan, P. (2006). An Australian historian puts Margaret Mead's biggest detractor on the psychoanalytic sofa. *The Chronicle of Higher Education*, 52(19): A14.

Money, J. (1985). *The Destroying Angel: Sex, Fitness & Food in the Legacy of Degeneracy Theory, Graham Crackers, Kellogg's Corn Flakes & American Health History*. Buffalo, NY: Prometheus Books.

———(2000, Fall). Wandering wombs and shrinking penises: The lineage and linkage of hysteria. *Link: A Critical Journal on the Arts in Baltimore and the World*, 5: 44–51.

Morgan, L. H. (1877/1908). *Ancient Society or Researches in the Lines of Human Progress from Savagery through Barbarism to Civilization*. Chicago: Charles H. Kerr & Company.

Morin, J. (1995). *The Erotic Mind: Unlocking the Inner Sources of Sexual Passion and Fulfillment*. New York: HarperCollins.

Morris, D. (1967). *The Naked Ape: A Zoologist's Study of the Human Animal.* New York: McGraw-Hill.

———(1981). *The Soccer Tribe.* London: Jonathan Cape.

———(1998). *The Human Sexes: A Natural History of Man and Woman.* New York: Thomas Dunne Books.

Moscucci, O. (1996). Clitoridectomy, circumcision and the politics of sexual pleasure in mid-Victorian Britain, in A. H. Miller and J. E. Adams (Eds.), *Sexualities in Victorian Britain.* Bloomington: Indiana University Press.

Moses, D. N. (2008). *The Promise of Progress: The Life and Work of Lewis Henry Morgan.* Columbia, MO: University of Missouri Press.

Namu, Y. E. (2004). *Leaving Mother Lake: A Girlhood at the Edge of the World.* New York: Back Bay Books.

Nishida, T., and Hiraiwa-Hasegawa, M. (1987). Chimpanzees and bonobos: Cooperative relationships among males. In B. B. Smuts, D. L. Cheney, R. M. Wrangham, and T. T. Struhsaker (Eds.), *Primate Societies* (pp. 165–177). Chicago: University of Chicago Press.

Nolan, P. D. (2003). Toward an ecological-evolutionary theory of the incidence of warfare in preindustrial societies. *Sociological Theory,* 21(1): 18–30.

O'Connell, J. F., Hawkes, K., Lupo, K. D., and Blurton Jones, N. G. (2002). Male strategies and Plio-Pleistocene archaeology. *Journal of Human Evolution,* 43: 831–872.

Okami, P., and Shackelford, T. K. (2001). Human sex differences in sexual psychology and behavior. *Annual Review of Sex Research.*

O'Neill, N., and O'Neill, G. (1972/1984). *Open Marriage: A New Life Style for Couples.* New York: M. Evans and Company.

Ostrom, E. (2009). A general framework for analizing sustainability of ecological systems. *Science,* 325: 419–422.

Parker, G. A. (1984). Sperm competition. In R. L. Smith (Ed.), *Sperm Competition and Animal Mating Systems.* New York: Academic Press.

Perel, E. (2006). *Mating in Captivity: Reconciling the Erotic and the Domestic.* New York: HarperCollins.

Pinker, S. (1997). Letter to the Editor of *New York Review of Books* on Gould. Retrieved January 22, 2002 from http://www.mit.edu/~pinker/GOULD.html.

———(2002). *The Blank Slate: The Modern Denial of Human Nature.* New York: Viking Press.

Pochron, S., and Wright, P. (2002). Dynamics of testis size compensates for variation in male body size. *Evolutionary Ecology Research,* 4: 577–585.

Pollock, D. (2002). Partible paternity and multiple maternity among the Kulina. In S. Beckerman and P. Valentine (Eds.), *Cultures of Multiple Fathers: The Theory and Practice of Partible Paternity in Lowland South America* (pp. 42–61). Gainesville: University Press of Florida.

Potts, M., and Short, R. (1999). *Ever since Adam and Eve: The Evolution of Human Sexuality.* Cambridge, UK: Cambridge University Press.

Potts, R. (1992). The hominid way of life. In Jones, S., Martin, R. D., and Pilbeam, D. (Eds.) (1992). *The Cambridge Encyclopedia of Human Evolution.* Cambridge, UK: Cambridge University Press, pp. 325–334.

Pound, N. (2002). Male interest in visual cues of sperm competition risk. *Evolution and Human Behavior,* 23: 443–466.

Power, M. (1991). *The Egalitarians: Human and Chimpanzee.* Cambridge, UK: Cambridge University Press.

Pradhan, G. R., et al. (2006). The evolution of female copulation calls in primates: A review and a new model. *Behavioral Ecology and Sociobiology,* 59(3): 333–343.

Prescott, J. (1975). Body pleasure and the origins of violence. *Bulletin of the Atomic Scientists,* November: 10–20.

Pusey, A. E. (2001). Of apes and genes. In F. M. de Waal (Ed.), *Tree of Origin: What Primate Behavior Can Tell Us About Human Social Evolution.* Cambridge, MA: Harvard University Press.

Quammen, D. (2006). *The Reluctant Mr. Darwin: An Intimate Portrait of Charles Darwin and the Making of His Theory of Evolution.* New York: Norton.

Quinn, D. (1995). *Ishmael: An Adventure of the Mind and Spirit.* New York: Bantam Books.

Raverat, G. (1991). *Period Piece: A Cambridge Childhood.* Ann Arbor, MI: University of Michigan Press.

Reid, D. P. (1989). *The Tao of Health, Sex & Longevity: A Modern Practical Guide to the Ancient Way.* New York: Simon & Schuster.

Richards, D. A. J. (1979). Commercial sex and the rights of the person: A moral argument for the decriminalization of prostitution. *University of Pennsylvania Law Review,* 127: 1195–1287.

Richards, M. P., and Trinkaus, E. (2009). Isotopic evidence for the diets of European Neanderthals and early modern humans. In press (published online before print August 11, 2009, doi: 10.1073/pnas.0903821106).

Ridley, M. (1993). *The Red Queen: Sex and the Evolution of Human Nature.* New York: Penguin.

———(1996). *The Origins of Virtue: Human Instincts and the Evolution of Cooperation.* New York: Viking.

———(2006). *Genome: The Autobiography of a Species in 23 Chapters.* New York: Harper Perennial.

Rilling, J. K., et al. (2002). A neural basis for social cooperation. *Neuron,* 35: 395–405.

Roach, M. (2008). *Bonk: The Curious Coupling of Sex and Science.* New York: Norton.

Roberts, S. C., et al. (2004). Female facial attractiveness increases during fertile phase of the menstrual cycle. *Proceedings Biological Sciences,* August 7; 271, 5: S270–S272.

Rodman, P. S., and Mitani, J. C. (1987). Orangutans: Sexual dimorphism in a solitary species. In B. B. Smuts, D. L. Cheney, R. M. Seyfarth, R. W. Wrangham, and T. T. Struthsaker (Eds.), *Primate Societies* (pp. 146–154). Chicago: University of Chicago Press.

Roney, J. R., Mahler, S.V., and Maestripieri, D. (2003). Behavioral and hormonal responses of men to brief interactions with women. *Evolution and Human Behavior,* 24: 365–375.

Rose, L., and Marshall, F. (1996). Meat eating, hominid sociality and home bases revisited. *Current Anthropology,* 37: 307–338.

Roughgarden, J. (2004). *Evolution's Rainbow: Diversity, Gender and Sexuality in Nature and People.* Berkeley: University of California Press.

———(2007). Challenging Darwin's Theory of Sexual Selection. *Daedalus,* Spring Issue.

———(2009). *The Genial Gene: Deconstructing Darwinian Selfishness.* Berkeley: University of California Press.

Rousseau, J. J. (1994). *Discourse Upon the Origin and Foundation of Inequality Among Mankind.* New York: Oxford University Press. (Original work published 1755.)

Rüf, I. (1972). Le 'dutsee tui' chez les indiens Kulina de Perou [The 'dutsee tui' of the Kulina Indians of Peru]. *Bulletin de la Société Suisse des Américanistes,* 36: 73–80.

Rushton, J. P. (1989). Genetic similarity, human altruism and group selection. *Behavioral and Brain Sciences,* 12: 503–559.

Ryan, C., and Jethá, C. (2005). Universal human traits: The holy grail of evolutionary psychology. *Behavioral and Brain Sciences,* 28: 2.

Ryan, C., and Krippner, S. (2002, June/July). Review of the book *Mean Genes: From Sex to Money to Food, Taming Our Primal Instincts. AHP Perspective,* 27–29.

Safron, A., Barch, B., Bailey, J. M., Gitelman, D. R., Parrish, T. B., and Reber, P. J. (2007). Neural correlates of sexual arousal in homosexual and heterosexual men. *Behavioral Neuroscience*, 121 (2): 237–248.

Sahlins, M. (1972). *Stone Age Economics*. New York: Aldine de Gruyter.

———(1995). *How "Natives" Think: About Captain Cook, for Example*. Chicago: University of Chicago Press.

Saino, N., Primmer, C. R., Ellegren, H., and Moller, A. P. (1999). Breeding synchrony and paternity in the barn swallow. *Behavioral Ecology and Sociobiology*, 45: 211–218.

Sale, K. (2006). *After Eden: The Evolution of Human Domination*. Durham, NC: Duke University Press.

Sanday, P. R. (2002). *Women at the Center: Life in a Modern Matriarchy*. Ithaca, NY: Cornell University Press.

Santos, P.S., Schinemann, J.A., Gabardo, J., Bicalho, Mda. G. (2005). New evidence that the MHC influences odor perception in humans: A study with 58 Southern Brazilian students. *Hormones and Behavior*. 47(4): 384–388.

Sapolsky, R. M. (1997). *The Trouble with Testosterone and Other Essays on the Biology of the Human Predicament*. New York: Simon & Schuster.

———(1998). *Why Zebras Don't Get Ulcers: An Updated Guide to Stress, Stress-related Diseases and Coping*. New York: W. H. Freeman and Company.

———(2001). *A Primate's Memoir: A Neuroscientist's Unconventional Life Among the Baboons*. New York: Scribner.

———(2005). *Monkeyluv: And Other Essays on Our Lives as Animals*. New York: Scribner.

Sapolsky R. M., and Share, L. J. (2004). A pacific culture among wild baboons: Its emergence and transmission. *PLoS Biology*, 4(2): e106. http://www.ncbi.nlm.nih.gov/pmc/articles/PMC387274/.

Savage-Rumbaugh, S., and Wilkerson, B. (1978). Socio-sexual behavior in *Pan paniscus* and *Pan troglodytes:* A comparative study. *Journal of Human Evolution*, 7: 327–344.

Scheib, J. (1994). Sperm donor selection and the psychology of female choice. *Ethology and Sociobiology*, 15: 113–129.

Schlegel, A. (1995). The cultural management of adolescent sexuality. In P. R. Abramson and S. D. Pinkerton (Eds.), *Sexual Nature / Sexual Culture*. Chicago: University of Chicago Press.

Schrire, C. (1980). An inquiry into the evolutionary status and apparent identity of San hunter-gatherers. *Human Ecology*, 8: 9–32.

Seeger, A., Da Matta, R., and Viveiros de Castro, E. (1979). A construção da pessoa nas sociedades indígenas brasileiras [The construction of the person in indigenous Brazilian societies]. *Boletim do Museu Nacional (Rio de Janeiro),* 32: 2–19.

Semple, S. (1998). The function of Barbary macaque copulation calls. *Proceedings in Biological Sciences,* 265(1393): 287–291.

———(2001). Individuality and male discrimination of female copulation calls in the yellow baboon. *Animal Behaviour* 61: 1023–1028.

Semple, S., McComb, K., Alberts, S., and Altmann, J. (2002). Information content of female copulation calls in yellow baboons. *American Journal of Primatology,* 56: 43–56.

Seuanez, H. N., Carothers, A. D., Martin, D. E., and Short, R. V. (1977). Morphological abnormalities in spermatozoa of man and great apes. *Nature,* 270: 345–347.

Seyfarth, R. M. (1978). Social relationships among adult male and female baboons: Behavior during sexual courtship. *Behaviour,* 64: 204–226.

Shackelford, T. K., Goetz, A. T., McKibbin, W. F., and Starratt, V. G. (2007). Absence makes the adaptations grow fonder: Proportion of time apart from partner, male sexual psychology and sperm competition in humans (*Homo sapiens*). *Journal of Comparative Psychology,* 121: 214–220.

Shaw, G. B. (1987). *Back to Methuselah.* Fairfield, IA: 1st World Library.

Shea, B. T. (1989). Heterochrony in human evolution: The case for neoteny reconsidered. *Yearbook of Physical Anthropology,* 32: 93–94.

Sherfey, M. J. (1972). *The Nature and Evolution of Female Sexuality.* New York: Random House.

Shores, M. M., et al. (2004). Increased incidence of diagnosed depressive illness in hypogonadal older men. *Archives of General Psychiatry,* 61: 162–167.

Shores, M. M., Matsumoto, A. M, Sloan, K. L., and Kivlahan, D. R. (2006). Low serum testosterone and mortality in male veterans. *Archives of Internal Medicine,* 166: 1660–1665.

Short, R. V. (1979). Sexual selection and its component parts, somatic and genital selection, as illustrated by man and the great apes. *Advances in the Study of Behavior,* 9: 131–158.

———(1995). Human reproduction in an evolutionary context. *Annals of New York Academy of Science,* 709: 416–425.

———(1998). Review of the book *Human Sperm Competition: Copulation, Masturbation and Infidelity.* Retrieved January 22, 2000 from http://wwwvet.murdoch.edu.au/spermology/rsreview.html.

Shostak, M. (1981). *Nisa: The Life and Works of a !Kung Woman.* New York: Random House.

———(2000). *Return to Nisa.* Cambridge, MA: Harvard University Press.

Siepel, A. (2009). Phylogenomics of primates and their ancestral populations. *Genome Research* 19: 1929–1941.

Singer, P. (1990). *Animal Liberation.* New York: New York Review Books.

Singh, D., and Bronstad, P. M. (2001). Female body odour is a potential cue to ovulation. *Proceedings in Biological Sci*ences, 268(1469): 797–801.

Small, M. F. (1988). Female primate sexual behavior and conception: Are there really sperm to spare? *Current Anthropology,* 29(1): 81–100.

———(1993). *Female Choices: Sexual Behavior of Female Primates.* Ithaca, NY: Cornell University Press.

———(1995). *What's Love Got to Do with It? The Evolution of Human Mating.* New York: Anchor Books.

Smith, D. L. (2007). *The Most Dangerous Animal: Human Nature and the Origins of War.* New York: St. Martin's Press.

Smith, J. M. (1991). Theories of sexual selection. *Trends in Ecology and Evolution,* 6: 146–151.

Smith, R. L. (1984). Human sperm competition. In R. Smith (Ed.), *Sperm Competition and the Evolution of Animal Mating Systems* (pp. 601–660). New York: Academic Press.

Smuts, B. B. (1985). *Sex and Friendship in Baboons.* New York: Aldine.

———(1987). Sexual competition and mate choice. In B. B. Smuts, D. L. Cheney, R. M. Seyfarth, R. W. Wrangham, and T. T. Struthsaker (Eds.), *Primate Societies* (pp. 385–399). Chicago: University of Chicago Press.

Sober, E., and Wilson, D. (1998). *Unto Others: The Evolution and Psychology of Unselfish Behavior.* Cambridge, MA: Harvard University Press.

Speroff, L., Glass, R. H., and Kase, N. G. (1994). *Clinical and Gynecologic Endocrinology and Infertility.* Baltimore, MD: Williams and Wilkins.

Sponsel, L. (1998). Yanomami: An arena of conflict and aggression in the Amazon. *Aggressive Behavior,* 24: 97–122.

Squire, S. (2008). *I Don't: A Contrarian History of Marriage.* New York: Bloomsbury USA.

Sprague, J., and Quadagno, D. (1989). Gender and sexual motivation: An exploration of two assumptions. *Journal of Psychology and Human Sexuality,* 2: 57.

Stanford, C. (2001). *Significant Others: The Ape—Human Continuum and the Quest for Human Nature.* New York: Basic Books.

Stoddard, D. M. (1990). *The Scented Ape: The Biology and Culture of Human Odour.* Cambridge, UK: Cambridge University Press.

Strier, K. B. (2001). Beyond the apes: Reasons to consider the entire primate order. In F. de Waal (Ed.), *Tree of Origin: What Primate Behavior Can Tell Us About Human Social Evolution* (pp. 69–94). Cambridge, MA: Harvard University Press.

Sturma, M. (2002). *South Sea Maidens: Western Fantasy and Sexual Politics in the South Pacific.* New York: Praeger.

Sulloway, F. (April 9, 1998). Darwinian virtues. *New York Review of Books.* Retrieved December 12, 2002 from http://www.nybooks.com/articles/894.

Symons, D. (1979). *The Evolution of Human Sexuality.* New York: Oxford University Press.

———(1992). On the use and misuse of Darwinism in the study of human behavior. In J. H. Barkow (Ed.), *The Adapted Mind: Evolutionary Psychology and the Generation of Culture* (pp. 137–159). New York: Oxford University Press.

Szalay, F. S., and Costello, R. K. (1991). Evolution of permanent estrus displays in hominids. *Journal of Human Evolution,* 20: 439–464.

Tanaka, J. (1987). The recent changes in the life and society of the central Kalahari San. *African Study Monographs,* 7: 37–51.

Tannahill, R. (1992). *Sex in History.* Lanham, MD: Scarborough House.

Tarín, J. J., and Gómez-Piquer, V. (2002). Do women have a hidden heat period? *Human Reproduction,* 17(9): 2243–2248.

Taylor, S. (2002). Where did it all go wrong? James DeMeo's Saharasia thesis and the origins of war. *Journal of Consciousness Studies,* 9(8): 73–82.

Taylor, T. (1996). *The Prehistory of Sex: Four Million Years of Human Sexual Culture.* New York: Bantam.

Testart, A. (1982). Significance of food storage among hunter-gatherers: Residence patterns, population densities and social inequalities. *Current Anthropology,* 23: 523–537.

Theroux, P. (1989). *My Secret History.* New York: Ivy Books.

Thompson, R. F. (1984). *Flash of the Spirit: African & Afro-American Art & Philosophy.* London: Vintage Books.

Thornhill, R., Gangestad, S. W., and Comer, R. (1995). Human female orgasm and mate fluctuating asymmetry. *Animal Behaviour,* 50: 1601–1615.

Thornhill, R., and Palmer, C. T. (2000). *A Natural History of Rape: Biological Bases of Sexual Coercion.* Cambridge, MA: The MIT Press.

Tierney, P. (2000). *Darkness in El Dorado: How Scientists and Journalists Devastated the Amazon*. New York: Norton.

Todorov, T. (1984). *The Conquest of America*. New York: HarperCollins.

Tooby, J., and Cosmides, L. (1990). The past explains the present: Emotional adaptations and the structure of ancestral environments. *Ethology and Sociobiology*, 11: 375–424.

———(1992). The psychological foundations of culture. In J. H. Barkow, L. Cosmides, and J. Tooby (Eds.), *The Adapted Mind: Evolutionary Psychology and the Generation of Culture* (pp. 19–136). Oxford, England: Oxford University Press.

———(1997). Letter to the Editor of *New York Review of Books* on Gould. Retrieved January 22, 2002 from http://cogweb.english.ucsb.edu/Debate/CEP_Gould.html.

Tooker, E. (1992). Lewis H. Morgan and his contemporaries. *American Anthropologist*, 94: 357–375.

Townsend, J. M., and Levy, G. D. (1990a). Effect of potential partners' costume and physical attractiveness on sexuality and partner selection. *Journal of Psychology*, 124: 371–389.

———(1990b). Effect of potential partners' physical attractiveness and socioeconomic status on sexuality and partner selection. *Archives of Sexual Behavior*, 19: 149–164.

Trivers, R. L. (1971). The evolution of reciprocal altruism. *Quarterly Review of Biology*, 46: 35–57.

———(1972). Parental investment and sexual selection. In B. Campbell (Ed.), *Sexual Selection and the Descent of Man* (pp. 136–179). Chicago: Aldine.

Turchin, P. (2003). *Historical Dynamics: Why States Rise and Fall*. Princeton, NJ: Princeton University Press.

Turchin, P., with Korateyev, A. (2006). Population density and warfare: A reconsideration. *Social Evolution & History*, 5(2): 121–158.

Turner, T. (1966). *Social Structure and Political Organization among the Northern Kayapó*. Unpublished doctoral dissertation, Harvard University, Cambridge, MA.

Twain, M. (1909/2008). *Letters from the Earth*. Sioux Falls, SD: Nu Vision Publications.

Valentine, P. (2002). Fathers that never exist. In S. Beckerman and P. Valentine (Eds.), *Cultures of Multiple Fathers: The Theory and Practice of Partible Paternity in Lowland South America* (pp. 178–191). Gainesville: University Press of Florida.

van der Merwe, N. J. (1992). Reconstructing prehistoric diet. In S. Jones, R. Martin, and D. Pilbeam (Eds.), *The Cambridge Encyclopedia of Human Evolution* (pp. 369–372). Cambridge, England: Cambridge University Press.

van Gelder, S. (1993). Remembering our purpose: An interview with Malidoma Somé. *In Context: A Quarterly of Humane Sustainable Culture,* 34: 30.

Ventura, M. (1986). *Shadow Dancing in the U.S.A.* Los Angeles: Jeremy Tarcher.

Verhaegen, M. (1994). Australopithecines: Ancestors of the African apes? *Human Evolution,* 9: 121–139.

Wade, N. (2006). *Before the Dawn: The Lost History of Our Ancestors.* New York: The Penguin Press.

Wallen, K. (1989). Mate selection: Economics and affection. *Behavioral and Brain Sciences,* 12: 37–38.

Washburn, S. L. (1950). The analysis of primate evolution with particular reference to the origin of man. Cold Spring Harbor Symposium. *Quantitative Biology,* 15: 67–78.

Washburn, S. L., and Lancaster, C. S. (1968). The evolution of hunting. In R. B. Lee and I. DeVore (Eds.), *Man the Hunter* (pp. 293–303). New York: Aldine.

Watanabe, H. (1968). Subsistence and ecology of northern food gatherers with special reference to the Ainu. In R. Lee and I. Devore (Eds.), *Man the Hunter* (pp. 69–77). Chicago: Aldine.

Wedekind, C., Seebeck, T., Bettens, F., and Paepke, A. J. (1995). MHC-dependent mate preferences in humans. *Proceedings of the Royal Society of London,* 260: 245–249.

——— (2006). The intensity of human body odors and the MHC: Should we expect a link? *Evolutionary Psychology,* 4: 85–94. Available online at http://www.epjournal.net/.

Weil, A. (1980). *The Marriage of the Sun and the Moon.* Boston: Houghton Mifflin.

White, T. D. (2009). *Ardipithecus ramidus* and the paleobiology of early hominids. *Science,* 326: 64, 75–86.

Widmer, R. (1988). *The Evolution of the Calusa: A Nonagricultural Chiefdom on the Southwest Florida Coast.* Tuscaloosa: University of Alabama Press.

Wiessner, P. (1996). Leveling the hunter: Constraints on the status quest in foraging societies. In P. Wiessner and W. Schiefenhovel (Eds.), *Food and the Status Quest: An Interdisciplinary Perspective* (pp. 171–191). Providence, RI: Berghahn.

Wilbert, J. (1985). The house of the swallow-tailed kite: Warao myth and the art of thinking in images. In G. Urton (Ed.), *Animal Myths and Metaphors in South America* (pp. 145–182). Salt Lake City: University of Utah Press.

Williams, G. C. (1966). *Adaptation and Natural Selection: A Critique of Some Current Evolutionary Thought.* Princeton, NJ: Princeton University Press.

Williams, W. L. (1988). *The Spirit and the Flesh: Sexual Diversity in American Indian Culture.* Boston: Beacon Press.

Wilson, E. O. (1975). *Sociobiology: The New Synthesis.* Cambridge, MA: The Belknap Press of Harvard University Press.

———(1978). *On Human Nature.* Cambridge, MA: Harvard University Press.

———(1998). *Consilience: The Unity of Knowledge.* New York: Knopf.

Wilson, J. Q. (2003). The family way: Treating fathers as optional has brought social costs. *The Wall Street Journal,* January 17, p. 7.

Wilson, M. L., and Wrangham, R. W. (2003). Intergroup relations in chimpanzees. *Annual Review of Anthropology,* 32: 363–392.

Wolf, S., et al. (1989). Roseto, Pennsylvania 25 years later—highlights of a medical and sociological survey. *Transactions of the American Clinical and Climatological Association,* 100: 57–67.

Woodburn, J. (1981/1998). Egalitarian societies. In J. Gowdy (Ed.), *Limited Wants, Unlimited Means: A Reader on Hunter-gatherer Economics and the Environment* (pp. 87–110). Washington, DC: Island Press.

Won, Yong-Jin, and Hey, J. (2004). Divergence population genetics of chimpanzees. *Molecular Biology and Evolution,* 22(2): 297–307.

World Health Organization. (1998). *Female Genital Mutilation: An Overview.* Geneva, Switzerland.

Wrangham, R. (1974). Artificial feeding of chimpanzees and baboons in their natural habitat. *Animal Behaviour,* 22: 83–93.

———(2001). Out of the *Pan,* into the fire: How our ancestors' evolution depended on what they ate. In F. de Waal (Ed.), *Tree of Origin: What Primate Behavior Can Tell Us About Human Social Evolution* (pp. 119–143). Cambridge, MA: Harvard University Press.

Wrangham, R., and Peterson, D. (1996). *Demonic Males: Apes and the Origins of Human Violence.* Boston: Houghton Mifflin.

Wright, R. (1994). *The Moral Animal: The New Science of Evolutionary Psychology.* New York: Pantheon.

Wyckoff, G. J., Wang, W., and Wu, C. (2000). Rapid evolution of male reproductive genes in the descent of man. *Nature,* 403: 304–308.

Yoder, V. C., Virden, T. B., III, and Amin, K. (2005). Pornography and loneliness: An association? *Sexual Addiction & Compulsivity,* 12: 1.

Zihlman, A. L. (1984). Body build and tissue composition in *Pan paniscus* and *Pan troglodytes,* with comparisons to other hominoids. In R. L. Susman (Ed.), *The Pygmy Chimpanzee* (pp. 179–200). New York: Plenum.

Zihlman, A. L., Cronin, J. E., Cramer, D. L., and Sarich, V. M. (1978). Pygmy chimpanzee as a possible prototype for the common ancestor of humans, chimpanzees and gorillas. *Nature,* 275: 744–746.

Zohar, A., and Guttman, R. (1989). Mate preference is not mate selection. *Behavioral and Brain Sciences,* 12: 38–39.

INDEX

Insights,
Interviews
& More . . .

About the authors

About the book

Read on

Meet Christopher Ryan and Cacilda Jethá

Philip Carr

Christopher Ryan received a BA in English from Hobart College in 1984 and an MA and PhD in psychology from Saybrook University in San Francisco, California, twenty years later. He spent the intervening decades in unexpected places working at very odd jobs (e.g., gutting salmon in Alaska, teaching English to prostitutes in Bangkok and self-defense to land-reform activists in Mexico, managing commercial real estate in New York's Diamond District, helping Spanish physicians publish their research). Drawing upon his multicultural experience, Christopher's doctoral research focused on trying to distinguish the human from the cultural by analyzing the prehistoric roots of human sexuality, and was guided by the world-renowned psychologist Stanley Krippner.

Based in Barcelona since the mid-1990s, Christopher has lectured at the University of Barcelona Medical

School and consulted at various local hospitals. He speaks to audiences around the world (in both English and Spanish). His work has appeared in major newspapers and magazines in many languages, scholarly journals, and a textbook used in medical schools and teaching hospitals throughout Spain and Latin America.

Christopher contributes blogs to both *Psychology Today* and *The Huffington Post.*

Cacilda Jethá has an Indian face, a European education, and an African soul. She was born in Mozambique to a mixed Muslim/Hindu family that had immigrated from Iran and India a few generations earlier. As a child, she fled civil war to Portugal, where she received most of her education and began her medical training before returning to Mozambique in the 1980s. As a young physician determined to help heal her country, Cacilda spent several years as the only physician serving some fifty thousand people in a vast rural district. While there, she conducted research on the sexual behavior of rural Mozambicans in order to help develop more effective AIDS-prevention efforts.

After almost a decade in Mozambique, Cacilda returned to Portugal, where she completed residency training in both psychiatry (at the prestigious Hospital Júlio de Matos in Lisbon) and occupational medicine. ▸

Meet Christopher Ryan and Cacilda Jethá
(continued)

She and Christopher live together in Barcelona, Spain, where she is a practicing psychiatrist. She speaks Portuguese, French, Spanish, Catalan, English, and some rusty Tsonga. ~

Christopher Ryan and Dan Savage in Conversation

The following is heavily edited from two Savage Love *podcasts recorded with Dan Savage in 2010 as well as an unpublished conversation. The full, unedited, raunchy, NSFW podcast recordings can be found (free) at the iTunes store or at www.thestranger.com. Episode 194 was broadcast on July 6, and episode 210 on October 27. Dan Savage is the best-known sex advice columnist in the United States. His podcast and syndicated column reach millions of readers weekly.*

Dan Savage: ***You know how every once in a while you see a movie or you read a book or an article in a magazine—or perhaps some advice in an advice column—and you think, "Oh my god, I'm not crazy!"? I've just had that experience reading your book. I don't know how to start this interview with you but by gushing. I understand that you're married to a woman, but where have you been all my life?***

Christopher Ryan: (laughing) I've been trying to get your attention for the past couple of years!

DS: ***I've been reading your book and screaming and yelling and jumping up and down. I feel so vindicated by the research that you guys have done because for years, I've been pointing*** ▸

out just from observation and anecdotal evidence that monogamy is hard, doesn't work, and makes people kind of miserable. And yet we're told it should come easily and naturally when we're in love, that a monogamous commitment is effortless where there's love. And you guys have demonstrated that this is simply not true.

CR: That's our conclusion.

DS: ***What inspired you to write this book, to paint this bull's-eye on your backs? Because you are going to get criticized from all sides.***

CR: I guess what got me into this line of thinking initially was the Clinton-Lewinsky situation. I was thinking, How is it possible that if men have had all the power—political, economic, even physical power—since the beginning of time, how is it that the most powerful man in the world is being publicly humiliated for having a consensual sexual relationship with someone? It just didn't make sense. So that led me to investigate evolutionary psychology. For the first year or so, I had the passion of the convert. I thought it explained everything. Luckily, at the time, I was living in San Francisco, working for a nonprofit organization called Women in Community Service. I was one of the only men working there. Consequently, I had lots of very intelligent, outspoken women around me who helped me see that the depiction of female sexuality that is fundamental to evolutionary psychology really doesn't make much sense at all.

DS: ***And what is the picture painted of female sexuality? You say in the book that "women are whores"—that's what they're telling us.***

CR: *We* don't say that! We half-jokingly say that *Darwin* says women are whores, in that they supposedly trade sex for stuff: protection, food, status, and so on, according to the conventional Darwinian view. We argue that women aren't whores by nature; they're sluts . . . and we mean that as a compliment! In other words, women evolved to have sex for the same reasons men do—because they like it. It feels good. Not because they're trying to get something from men.

DS: ***There's stuff in this book that will make people's heads explode, like the concept of partible paternity. Instead of women trading sexual exclusivity and the assurance of paternity to men, that what was actually going on was that women may have been fucking as many men as possible so all the men in the group could believe that the kid could be theirs.***

CR: Right. This is something that Sarah Hrdy writes about in several of her books on alloparenting. She explains that it's actually better for that child if lots of adults feel they have a direct connection to the child.

DS: ***You guys argue that this was the natural order of things, sort of a polyamorous, all-for-one-and-one-for-all sex culture. So, if monogamy wasn't our natural state, and now it's an "ill-fitting garment," and so on, now what?***

CR: Well, this is a book about how we are and how we got to be this way, but it's not about what anyone should do about it.

DS: ***Yes, but don't you think it will help people just to know the reason that they're falling short? That they're not doing anything wrong, necessarily? That it's their inner nature, their inner bonobo? These ideals our culture has created about monogamy and faithfulness and fidelity are nice, but they're not very functional and our own libidos and reptile brains are at war with them.***

CR: Exactly. People often say to us, "But we're human beings. We can *choose* how to behave." That's true, to a certain extent, but our bodies rebel against decisions that go against our evolved nature. You can *choose* to wear shoes that are too small, but you can't choose to be comfortable in them. You can *choose* to wear a corset, but you may well pass out because you can't breathe ▸

properly. . . . The human body and mind have evolved for a certain kind of life. The further we diverge from that path, the greater the cost in terms of mental, emotional, and physical health. There's just no getting away from this. We examine all this in greater detail in our next book.

Getting back to your earlier question—"Now what?"—the ambition we have for this book is humble but important, I think. We hope it encourages and empowers people to give themselves a break, to cut themselves and their partners some slack. It actually promotes family stability to not be so rigid concerning fidelity. A zero-tolerance policy doesn't help anyone.

DS: ***I'm constantly arguing that! If we're interested in preserving marriages and keeping families intact, we need to be less psychotic about never seeing anyone else naked ever again once we're married. If we make certain allowances, the marriage is more likely to survive over the long run. If the only way you can ever have sex with anyone else is to end the marriage, many people will end or sabotage the marriage.***

CR: Well put.

DS: ***OK, let's talk about how the book's been received so far. I've seen some negative reaction concerning jealousy and love. What do you say to people who say this vision of prehistory can't possibly be right because we're "by nature" possessive about love?***

CR: Well, I generally avoid making sweeping statements about these issues, but I'll give it a shot for you, Dan. Cacilda and I believe that real love isn't primarily about sex. If you're lucky enough to find love in your life, you quickly realize how relatively unimportant sex is. Love is about a lot of unerotic things: getting old together, taking care of each other when we're sick or grieving, raising a child together, paying the bills . . . sharing the dailyness of life. It's not primarily about orgasms. So many people confuse these things. They mistake good sexual chemistry for soul-mating, for a reason to sign up for a life together. Then, a few years later, when the chemical thrill has dissipated—as it does—they find they've made a horrible mistake. So, even though our book is largely about sex, one of the central points we wanted to make is that most of us take sex way too seriously. We need to chill out. Like music, sex *can* be sacred but doesn't always *have* to be. Sometimes we hear God in a Bach toccata, but sometimes we're just dancing and having a good time listening to the Rolling Stones. Nothing sacred about it.

DS: ***You heard it here first, folks! Other than those quibblers, how's the book been received so far?***

CR: To be honest, we're surprised and thrilled at the reception. Of course, I've got a Google alert set to the title, so I see most of the blog discussions, reviews on Amazon.com and GoodReads, etc. Overall, people seem to be very positive about the book. Much more than we'd expected. We figured we'd get about 50 percent hate mail and 50 percent non-hate mail, but probably 90 percent of the people who've written to us have been very, very kind in their comments. Some of the emails are heartbreaking. A lot of them along the lines of, "Damn, I wish I'd known all this when I was young!" One I'll never forget said, simply, "I'm a sixty-three-year-old widow and I consider this one of the most important books I've ever read. I wish I could live my life over with this information." We get a lot of messages that relate to what you said earlier, about how people feel better when they can understand why they feel as they do. Why they can honestly love their partner but still feel attracted to other people. Seriously, the response from readers has been mind-blowing.

DS: ***But surely not* all *positive.***

CR: No, and maybe this is still the calm before the storm. There hasn't been much negative response from the academic community—at least not the evolutionary psychology crowd we critique in our book: Pinker, Buss, Chagnon, and those guys.

We've received lovely messages from people like Frans de Waal and Sarah Hrdy—certainly two of the most prominent authorities in the evolution of human sexuality—although ▸

they don't necessarily agree with everything we wrote. On the other hand, Alan Dixson, a prominent primatologist we quote extensively, trashed us in an interview he gave to a paper in New Zealand, but he admitted he hadn't read the book, he was just responding to what he'd heard about it. That's been sort of typical of the most negative critical responses so far. They admit they haven't read *Sex at Dawn*, but they dismiss it anyway. Megan McArdle, the business and economics editor at *The Atlantic*, wrote what is probably the most negative review to date, and she openly admitted she was only halfway through the book when she wrote it.

DS: ***I saw that! "Humans aren't like bonobos . . . because we're not like bonobos!"***

CR: Right. Rock-solid logic there, no? She apparently felt so threatened by the book that she couldn't think straight. I mean, she accused us of leaving out any discussion of jealousy as it didn't fit our model, somehow missing Chapter 10, which is called "Jealousy: A Beginner's Guide to Coveting Thy Neighbor's Spouse."

But as I said, the negative response has been much less than we were expecting, and is understandable; lots of people feel threatened by the arguments we've made.

We've received lots of support from academics and clinicians. We've heard from scholars at the Kinsey Institute, professors all over the country who are assigning the book to their students, and we were honored when *Sex at Dawn* was chosen as the best consumer book of 2010 by the Society for Sex Therapy and Research (SSTAR).

DS: ***I'm so sorry Cacilda couldn't join you on this trip. What was her role in the book?***

CR: Cacilda is one of two psychiatrists who run a psychiatric facility with close to a hundred patients. So she's pretty tied to her day job these days, much as she'd love to participate more in interviews and meet readers. Strangely, a few people have interpreted her non-presence in interviews as evidence that I made her up just to give the book some cover with women readers. Seriously!

I'd done a lot of this research before meeting Casi, about ten years ago. Still, what she brought to the book was crucial. First, she'd done her own research into human sexual behavior in rural Mozambique for the World Health Organization in the first years of the AIDS crisis, so she had extensive "real life" understanding of how things work in that part of the world. She grew up in a mixed Muslim/Hindu Indian family in Africa, so she brought a lot of multicultural nuance to the project, and of course, being a medical doctor, she was integral to the discussions of diet, longevity, infant care, and so on. Portuguese is her native language, and English is actually one of six that she speaks. As the native English speaker and professional writer, I did the writing, but she read every draft, again and again, before it went to anyone else. To call her anything other than a coauthor would be inadequate.

DS: ***You've lived in Spain for a long time. Have you seen any major differences in the way Americans and Spanish deal with sex?***

CR: Oh yeah. In fact, that may be *why* I've lived in Spain for so long! Despite the history of Catholicism as the "official religion," urban Spanish people, at least, are far less conflicted about sex than the typical American. One of the first things that struck me was how openly and unashamedly Spanish people flirt. I'm not, and never have been, a particularly great-looking guy, but after a few weeks walking around in Barcelona, I felt like Brad Pitt! It's not sleazy or even necessarily sexual, but women look in your eyes, and if they like what they see, they smile. So simple. There's not that fear and suspicion of strangers one finds so often in American cities, where eye contact is to be avoided at all costs. In the U.S., there seems to be an assumption that any man you don't know could very well be a rapist-pervert-murderer or creep of some sort. I'm not blaming women for having that fear, of course, but it's pretty depressing for men *and* women. In ▸

Barcelona, you can walk down the street and be smiled at by three or four lovely women per block. It sure makes walking a lot more fun!

Cacilda gets the same sort of ego boosts all the time. (But she *is* particularly beautiful, so it's less surprising.) Nothing sleazy, mind you, but just guys who say things like, "Hi. I just wanted to introduce myself and ask if you'd like to have a drink sometime. You're really lovely." There's an innocence around flirting here that's been lost in the States. It's a shame, as it's very much a win-win situation that dramatically improves quality of life for everyone involved.

DS: ***What was the most surprising thing you learned while working on* Sex at Dawn*?***

CR: So many things come to mind, but probably the most mind-blowing was the information about the first "key parties" having been started by elite WWII pilots who were facing very high fatality rates, the highest in the whole military. It was so moving to think about what motivated them to open their marriages with other couples. They were cultivating these webs of love, or at least real affection, because they knew that some of the men wouldn't survive the war, and they wanted the widows to have as much support and love as possible. This confluence of selflessness and sexuality seemed to connect so directly to the hunter-gatherer groups, where men also have a high mortality rate from hunting accidents, falls, animal attacks, and so on. It was an unexpected yet very clear reflection of the distant past.

DS: ***Has the experience of publishing this book changed you in any way?***

CR: Interesting question. It has. It's made me much less critical of other people's books. I don't think I could write a negative review at this point. When I was younger, it was easy to point out the flaws in books, but at this point, I'm much more aware that there's a person on the other side who did their best. If I were writing

Sex at Dawn again, I'd probably tone down some of the snarkier bits.

I'm also more aware of just how impossible it is to please everyone. We've been incredibly fortunate in the response to this book, but still, for every nine comments we get congratulating us for writing the book with humor, we'll get one or two describing the writing style as "sophomoric" or "unserious." Some people think serious issues can only be discussed in serious tones. It's interesting to have become a public figure, even in the very limited way I'm experiencing it. People have the right to their responses to your work, positive or negative. I've learned not to take it personally, in either case. ~

Read on

Sex at Dawn Online

For more information about *Sex at Dawn* and the authors' current doings, please visit sexatdawn.com, where you'll find a selection of reviews, reader responses, TV and radio interviews, podcasts, and a reader-maintained forum for discussing anything and everything related to the book. Additionally, the authors maintain a Facebook page with a lively discussion of current events related to the book at www.facebook.com/sexatdawn. They can also be followed on Twitter: @SexatDawn.